THE LANGUAGE REVOLUTION

How Natural Language Processing is Powering Modern AI

Julio C. Enríquez & Sofía V. Enríquez

First Edition: 2026

ISBN: 979-8-9958716-0-6 (Paperback)
ISBN: 979-8-9958716-1-3 (Hardcover)
ISBN: 979-8-9958716-2-0 (Ebook)

Published by Julio C. Enriquez Press
Printed in the United States of America

Prologue

The Language Revolution

The world is in the midst of a technological transformation so profound that it is often compared to the industrial and internet revolutions. At the heart of this new era is artificial intelligence, a force that is reshaping industries, redefining professions, and fundamentally altering our interaction with the digital world. Yet, this revolution is not being driven by code alone. It is being driven by something far more human: language.

From the complex legal contracts that govern our economies to the billions of social media posts that shape our culture, the vast majority of human knowledge is stored as unstructured text. For decades, this universe of information remained largely opaque to computers. Today, that has changed. The ability to teach a machine to read, understand, and reason over human language—the core mission of natural language processing (NLP)—has unlocked the true potential of AI. NLP is the bridge between human thought and machine intelligence. It is the essential component that allows an AI to analyze a market research report, a doctor's note, or a customer review with a speed and scale that was once unimaginable.

Understanding NLP is, therefore, no longer a niche academic pursuit. It has become a fundamental prerequisite for understanding artificial intelligence itself. For any business that wants to implement a meaningful AI strategy, for any leader who wants to drive innovation, and for any professional who wants to remain relevant in the coming decade, fluency in the concepts of NLP is not optional. It is the new lingua franca of technology and business. The ability to grasp how a machine processes language will be the difference between leading the next wave of innovation and risking obsolescence in a job market that will increasingly demand AI literacy.

This book was written to provide that fluency.

Who Is This Book For?

This book is designed for an audience of builders, thinkers, and leaders. Its primary audience includes university students pursuing bachelor's, master's, or graduate-level degrees in Computer Science, Computer Engineering, and artificial intelligence. Furthermore, its focus on foundational concepts makes it equally relevant to business leaders, data analysts, and professionals in any field who recognize the need to understand the mechanics behind the AI tools that are reshaping their domains. The content is structured to be taught as a comprehensive, one-semester course at the university level or to be used as a self-study guide for any professional eager to master the foundations of NLP.

What This Book Is, and What It Is Not

The world is not short on NLP books. There are hundreds of excellent texts dedicated to the intricacies of programming with NLP libraries, and numerous others that delve into the deep, theoretical complexities of linguistics, dissecting sentences over hundreds of pages.

This book is not one of them.

This is not a book about coding. You will not find endless pages of Python scripts or library tutorials. We believe that a deep, conceptual understanding must precede implementation.

This is not a book of pure linguistics. While we honor the linguistic principles that underpin the field, our focus is on the computational models that make language processing possible at scale.

Instead, this is a foundational book with a business-oriented focus. Our goal is to build a powerful mental model of how NLP works, from the ground up. We will focus on the *why* and the *how* of the core algorithms, the architectural evolution of the models, and the real-world applications that drive business value. We will trace the journey from classic statistical methods to the state-of-the-art transformer architectures, providing you with the comprehensive knowledge needed to think strategically and critically about applying NLP in any context.

A Guide to the Journey

To achieve this, the book is organized into four distinct parts, each building upon the last to create a complete picture of the field.

- **Part 1: Foundations of Language and AI.** We begin by establishing the big picture, exploring the fundamental challenges of language, its relationship to AI, and the layers of analysis—from syntax to semantics—that form the theoretical bedrock of NLP.

- **Part 2: Core NLP Techniques and Building Blocks.** Here, we roll up our sleeves and dive into the classic, foundational techniques that are the prerequisites for any NLP task. We will cover everything from preprocessing and N-grams to the statistical models that powered the first wave of the NLP revolution.

- **Part 3: The Deep Learning Revolution.** This section tells the story of modern NLP. We will trace the evolution of neural architectures, from the first sequence-to-sequence models, through the critical invention of the attention mechanism, and culminating in the transformer—the architecture that powers nearly all modern AI.

- **Part 4: Applications and Advanced Topics.** In the final section, we put our knowledge to work. We will explore how these powerful techniques are applied to solve real-world problems in text analysis, information extraction, and social business intelligence. We will conclude with a dedicated look at the critical field of evaluating NLP systems and a crucial, capstone chapter on the ethical and societal implications of this powerful technology.

The journey ahead is both challenging and rewarding. By the end of this book, you will not just know the definitions of NLP terms; you will understand the narrative of a field in constant motion, and you will be equipped with the foundational knowledge to participate in its future. Welcome to the language of a revolution.

For a consolidated, comparative summary of this evolutionary path, readers may refer to the table in **Appendix G**, which provides a high-level overview of the progression of NLP paradigms from a computational and applied perspective.

About the Authors

Julio C. Enríquez is an Electrical Engineer with +25 years of professional experience in the Information and Communications Technology (ICT) industry. He has vast international experience working on high-complexity technological projects. Currently, he is leading global transformational projects at SAP, an enterprise application and business AI company. He also has an outstanding academic trajectory, holding several Master's Degrees in technology and business in the areas of Data Network Communication, Quality Systems, Business Administration, and Knowledge Management. He is currently pursuing a Doctoral Degree in Strategy and Artificial Intelligence.

Sofía V. Enríquez is a Computer Engineer from Villanova University with extensive experience in developing AI and NLP solutions for industry-leading organizations, specializing in the practical application of Transformer-based architectures. Currently, she is working as a Cloud Support Engineer at Strategy, an AI-powered enterprise analytics company.

Table of Contents

Part I

Foundations of Language and AI

Chapter 1

Chapter 2

Chapter 3

Part II

Core NLP Techniques and Building Blocks

Chapter 4

Chapter 5

N-grams: From Foundations to the Transformer Era 69

Chapter 6

Part-of-Speech Tagging 83

Chapter 7

The Penn Treebank Tagset: A Foundation for Computational Linguistics 105

Part III
The Deep Learning Revolution

Chapter 8

Foundations in Classic Machine Learning for NLP 119

Chapter 9

Machine Translation: The Quest for Meaning via Intermediate Representation 135

Chapter 10

Sequence-to-Sequence Models 147

Part IV
Applications and Advanced Topics

Chapter 14

Text Analysis in the Age of Artificial Intelligence...201

Chapter 15

A Guide to Evaluation in NLP 213

Chapter 16

Information Extraction: From Raw Text to Actionable Knowledge 229

Chapter 17

Applying Natural Language Processing for Social Business Intelligence 241

Chapter 18

Appendices

Appendix A

Appendix B

Appendix C

Appendix D

Appendix E

Appendix F

Appendix G

Appendix H

Appendix I

PART I

Foundations of Language and AI

Chapter 1

Natural Language Processing

1.1 Introduction to Natural Language Processing

Imagine asking your home assistant, "Play some upbeat jazz, but not anything from the last decade, and make it loud enough for a party." In a matter of seconds, it parses your complex request—discerning intent, musical genre, temporal constraints, and volume—and fills the room with sound. This seemingly magical interaction is powered by natural language processing (NLP), a vibrant and rapidly evolving field at the intersection of computer science, linguistics, and artificial intelligence. The primary mission of NLP is to build systems that can understand, interpret, manipulate, and even generate human language, bridging the vast chasm between our nuanced, often ambiguous communication and the rigid logic of a machine.

This chapter will guide you through the foundational concepts of NLP. We will begin by exploring the field's deep-rooted connections to linguistics and the core challenge that shapes our work: the pervasive nature of linguistic ambiguity. We will then trace the evolution of the models and algorithms designed to solve this puzzle, from early rule-based systems to the statistical methods that dominated for decades. Following this, we will dive into the current paradigm shift driven by the immense power of large language models (LLMs) and the underlying architectures that enable them. Finally, we will situate these advancements within the rich history of the field and look toward the ultimate goal of natural language understanding (NLU). By the end, you will not only grasp the core principles of NLP but also appreciate the intellectual journey that continues to push the boundaries of what machines can comprehend.

1.2 NLP Connections and Foundations

NLP is an interdisciplinary field at its core, deeply intertwined with computational linguistics, which provides the theoretical bedrock for understanding language structure, grammar, semantics, and pragmatics (Chomsky, 1957; Dikken, 2019). This linguistic knowledge is indispensable for designing algorithms that can untangle the complexities of human language. Furthermore, NLP applies to both written text and spoken language, demonstrating its independence from modality. While the initial digital representations of a sound wave and a text document differ, the underlying principles of understanding often converge. Techniques developed for text can often be adapted for speech after transcription. Consequently, comprehensive NLP systems must integrate components like speech recognition (audio-to-text) and text-to-speech synthesis (text-to-audio), a challenge that has been central to the field for decades (Rabiner & Juang, 1993). This fundamental reliance on linguistic theory and signal processing underscores the field's collaborative nature.

1.2.1 The Challenge of Ambiguity in NLP

The central, exhilarating puzzle that NLP researchers have been trying to solve for decades is ambiguity. Unlike formal programming languages, human language is rife with it (Jurafsky & Martin, 2023). This ambiguity manifests at multiple levels:

- **Lexical Ambiguity:** A single word can have multiple meanings. The word 'bank' can refer to a financial institution or the edge of a river. Without context, a machine is lost. This problem, known as polysemy, is a classic challenge in word sense disambiguation (Navigli, 2009).

- **Syntactic Ambiguity:** The grammatical structure of a sentence can be parsed in more than one way. Consider the classic example: "He saw a man on a hill with a telescope." The prepositional phrase "with a telescope" could modify 'man' (the man has the telescope) or 'saw' (he used the telescope to see the man). This structural ambiguity requires sophisticated parsing techniques to resolve (Manning & Schütze, 1999).

- **Semantic and Pragmatic Ambiguity:** The overall meaning can shift based on context and implied intent. If someone says, "The presentation was brilliant," their tone and the surrounding conversation determine whether it's sincere praise or biting sarcasm. This requires an understanding of pragmatics, the study of how context contributes to meaning (Grice, 1975;

Berstler, 2024; Hu, 2024).

Addressing ambiguity is not merely an academic exercise; it directly affects the performance of real-world NLP applications and is a central theme we will return to throughout this book.

The Concrete Impact of Ambiguity on NLP Systems

To truly appreciate why ambiguity is the central challenge of NLP, it is crucial to understand how it causes real-world systems to fail. Each type of ambiguity can lead to a distinct and critical error in a downstream application, demonstrating the necessity of contextual understanding (Dale, 2021).

Lexical Ambiguity's Impact on Machine Translation: Consider the sentence, "The company's stock is trading at a new peak." A machine translation system must resolve the lexical ambiguity of the word 'stock.' If it incorrectly chooses the meaning related to inventory or supply, it might translate the sentence into Spanish as "El *inventario* de la empresa está en un nuevo pico," which is nonsensical. A context-aware system, however, would recognize financial terms like 'company' and 'trading' and correctly select the financial meaning, producing the accurate translation: "La *acción* de la empresa cotiza en un nuevo máximo." This single-word disambiguation is the difference between a useless translation and an accurate one. The system's ability to perform word-sense disambiguation is fundamental to its utility (Yarowsky, 1995).

Syntactic Ambiguity's Impact on Question-Answering: Let's revisit the sentence "He saw a man on a hill with a telescope." Imagine this sentence is part of a document fed into a question-answering (QA) system. If a user asks, "How was the man seen?" the system's response depends entirely on its parsing of the sentence's syntax. If it attaches "with a telescope" to the verb 'saw,' it will correctly answer, "He was seen using a telescope." However, if it incorrectly attaches the phrase to 'man,' it might erroneously answer, "The man he saw was holding a telescope," which fails to answer the user's question. For a QA system to be reliable, it cannot simply match keywords; it must accurately model the grammatical relationships between them to understand who did what to whom, and with what tools.

Pragmatic Ambiguity's Impact on Virtual Assistants: A user says to their home assistant, "Wow, it's really cold in here." A system operating on literal semantic meaning might respond with a factual statement like, "The current temperature is 62 degrees Fahrenheit." While correct, this response is unhelpful because it misses the

Figure 1.1

The Three Levels of Ambiguity

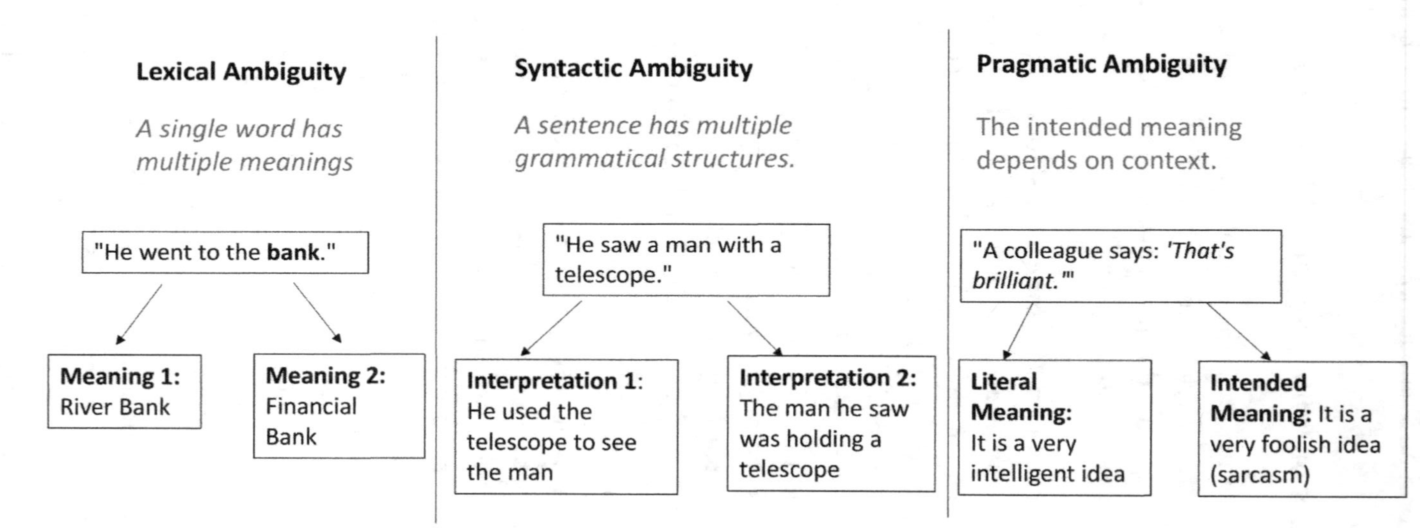

user's pragmatic intent. The user is not asking for a weather report; they are implicitly making a request. A system capable of pragmatic analysis would understand this implied intent and perform the desired action, responding with something like, "Okay, I'm raising the thermostat to 70 degrees." This ability to infer intent beyond the literal words is the hallmark of a truly intelligent and useful assistant (Cohen & Perrault, 1979; Dong et al., 2023; Kim et al., 2022).

1.2.2 Models, Algorithms, and Techniques in NLP

To tackle the challenge of ambiguity, the field has developed an increasingly sophisticated toolkit. The evolution of these techniques can be broadly categorized into three paradigms (Hirschberg & Manning, 2015):

- **Symbolic/Rule-Based Systems (1950s-1980s):** Early approaches involved manually codifying linguistic knowledge into explicit if-then rules. These systems were brittle, labor-intensive to build, and struggled to handle the near-infinite variation of natural language (Winograd, 1972).

- **Statistical NLP (1990s-2010s):** With the rise of larger datasets and greater computational power, the focus shifted to statistical methods and machine learning. Instead of being explicitly programmed, models could learn linguistic patterns from data. This data-driven approach produced more robust and scalable systems (Manning & Schütze, 1999).

- **Neural NLP (2010s-Present):** The deep learning revolution brought artificial neural networks to the forefront. These models, inspired by the structure of the human brain, can learn complex, hierarchical patterns in language data, leading to state-of-the-art performance on nearly every NLP task (Goldberg, 2017).

Before any of these models can be applied, raw text must first be cleaned and structured into a format a machine can understand. This crucial first step, known as **preprocessing**, is a foundational pipeline of techniques that serves as the bridge between unstructured human language and the structured numerical data required by machine learning algorithms. This pipeline involves several key stages. First, **tokenization** breaks a stream of text into its constituent words or sub-words. Next, **stop word removal** filters out common, low-information words (e.g., 'the', 'a', 'is') that can add noise to the analysis. Finally, techniques like **stemming** or **lemmatization** reduce words to their root or dictionary form (e.g., 'running', 'ran', and 'runs' all become 'run'). The collective goal of this pipeline is to reduce the complexity and dimensionality of the text data, creating a cleaner, more consistent, and computationally efficient representation. While the specific steps can vary depending on the task and the model being used, a well-designed preprocessing pipeline is absolutely critical for the success of any NLP project, a topic we will explore in extensive detail in Chapter 4.

1.3 The Impact of Large Language Models (LLMs)

The current state of NLP is dominated by large language models (LLMs). These are neural networks of unprecedented scale, trained on colossal amounts of text data, enabling them to acquire a broad range of world knowledge and linguistic capabilities (Brown et al., 2020). The watershed moment for LLMs was the introduction of the transformer architecture and its key innovation, the attention mechanism. This mechanism allows the model to dynamically weigh the importance of different words in the input when processing and generating language, effectively giving it a powerful form of contextual awareness (Vaswani et al., 2017). Models based on this architecture, from foundational systems like Google's BERT (Devlin et al., 2019) and OpenAI's initial GPT series (Min et al., 2024; Naik et al., 2024) to more recent large-scale models such as Meta's Llama series (Touvron et al., 2023), Google's Gemini (Gemini Team, 2023), and Anthropic's Claude family (Anthropic, 2024), have achieved remarkable performance.

1.3.1 Core Challenges of Large Language Models

However, this power is not without its challenges. The very nature of how LLMs are trained gives rise to inherent and significant risks that practitioners must understand and mitigate. Two of the most critical challenges are hallucination and the perpetuation of societal biases (Bender et al., 2021).

The Hallucination Problem: LLMs are trained to be masterful predictors of the next most plausible word in a sequence. They are not, however, connected to a real-time, fact-checked knowledge base. This can lead them to 'hallucinate'—generating text that is fluent, confident, and grammatically perfect, but factually incorrect. For example, a user might ask a well-known historical question: "In what year did Neil Armstrong walk on the moon?" An LLM might hallucinate a minor but critical detail, responding: "Neil Armstrong, the first person to walk on the moon, made his historic steps in 1968." The correct year is 1969. The model's response is plausible and well-formed, but it is a factual error. For applications in sensitive domains like medicine, law, or finance, this tendency to generate convincing falsehoods makes the raw output of LLMs unreliable without robust fact-checking or grounding mechanisms, a topic we will explore in the context of Responsible AI in Chapter 18.

Perpetuation of Societal Bias: Because LLMs learn from vast amounts of text from the internet, they inevitably absorb the stereotypes and biases present in that data. This learned bias can then surface in the model's outputs in subtle but harmful ways. For instance, if a model is given the prompt "The CEO delivered a speech to.. ." it might be statistically more likely to complete the sentence with "...his employees." Conversely, if given the prompt "The kindergarten teacher read a story to...", it may be more likely to complete it with "...her students." While seemingly innocuous, this reflects and reinforces societal gender stereotypes about professions. When these models are used in downstream applications such as resume screening or loan application analysis, these biases can lead to discriminatory and unfair outcomes. Mitigating this learned bias is one of the most pressing technical and ethical challenges in AI today (Bolukbasi et al., 2016).

1.4 NLP and Natural Language Understanding (NLU)

While NLP encompasses the entire spectrum of language processing, the subfield of natural language understanding (NLU) represents its ultimate ambition. NLU is focused on a much deeper challenge: achieving human-like comprehension. This means moving beyond surface-level pattern matching to interpret intent, resolve contextual ambiguity, and make logical inferences (Allen, 1995; McShane, 2017). The distinction is subtle but profound. An NLP system might correctly classify a movie review as 'negative', but an NLU system would aspire to understand why it's negative (e.g., "the plot was predictable, but the cinematography was stunning"). This deeper level of analysis is often referred to as "NLU-complete" and remains a grand challenge for the field, a point that remains relevant even in the era of large language models (Shadbolt et al., 2006; Bender et al., 2021).

1.4.1 A Deeper Example of the NLP vs. NLU Distinction

To fully grasp the difference between surface-level processing and deep understanding, consider a complex, multi-intent request a user might make to a sophisticated virtual assistant:

> "Find me a good Italian restaurant downtown that's not too expensive and has an open table for two tonight."

A system with strong **NLP** capabilities would successfully perform the initial, crucial steps of the analysis. It would use named entity recognition (NER) to identify key entities: 'Italian' (Cuisine Type), 'restaurant' (Place Type), 'downtown' (Location), and 'tonight' (Date/Time). It could parse the sentence structure to understand the relationships among these entities and extract keywords such as "table for two." This is a complex and valuable process of structuring the query (Nadeau & Sekine, 2007).

However, a system with true **NLU** capabilities must go significantly further, tackling the ambiguous and subjective parts of the request that require world knowledge and reasoning. It would need to:

1. **Interpret Subjectivity:** It must understand that 'good' is not an objective attribute. To resolve this, it would need to access an external knowledge base of restaurant reviews and ratings, inferring that 'good' likely means a rating of 4 stars or higher.

2. **Resolve Ambiguity:** It must define "not too expensive." This requires contextual knowledge of restaurant pricing in the 'downtown' area, perhaps by mapping the phrase to a specific price tier, such as '$$.'

Figure 1.2

NLP vs. NLU: The Restaurant Test

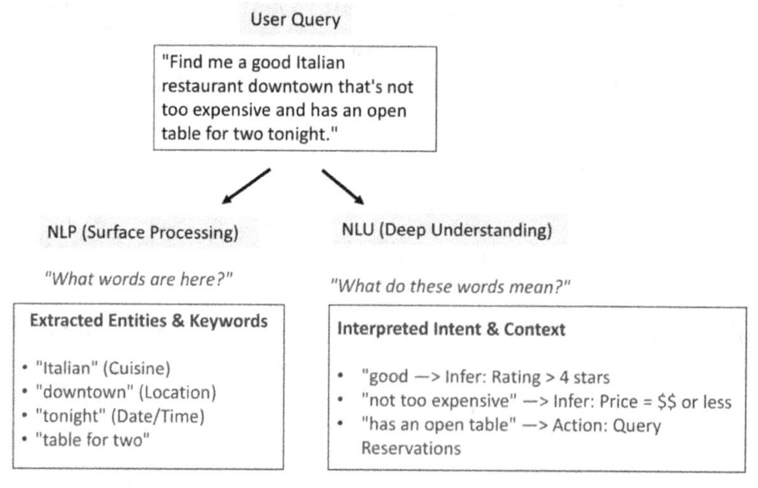

3. Synthesize and Act: Finally, it must synthesize all these structured and interpreted constraints—Cuisine: Italian, Location: Downtown, Rating: >4 stars, Price: <=$$, Availability: Table for 2, Date: Tonight—and execute a complex series of actions, such as querying a reservation database and presenting the user with a list of qualifying options.

This example highlights the chasm between processing language and truly understanding it. NLP provides the structured data, but NLU provides the reasoning and world knowledge needed to turn that data into a meaningful action (Bates, 1995; Ohmer et al., 2024).

1.5 A Brief History of NLP

The journey of NLP is one of ambitious goals, periods of disillusionment, and stunning breakthroughs. A timeline helps to visualize this evolution (Hutchins, 1995; Jurafsky & Martin, 2023):

- **1950s: The Dawn of an Idea.** The Georgetown-IBM experiment demonstrated automatic translation, sparking immense excitement.

- **1966: The ALPAC Report.** A pessimistic report on machine translation led to a significant funding cut and the first "AI winter."

- **1980s: The Rise of Expert Systems.** Rule-based systems saw a resurgence in highly specific domains.

- **1990s: The Statistical Revolution.** The field shifted dramatically towards machine learning and statistical methods.

- **2010s: The Deep Learning Revolution.** Deep neural networks began to achieve incredible results, becoming the standard for leading-edge NLP (Sutskever et al., 2014).

- **2017-Present: The Transformer Era.** The introduction of the transformer architecture (Vaswani et al., 2017) ushered in the age of LLMs, leading to the rapid advancements we see today.

1.6 Conclusion

Natural language processing is one of the most critical and fastest-moving disciplines in technology today. Its objective—to enable computers to understand and generate human language—is simple to state but profoundly complex to achieve. The journey from fragile, rule-based systems to massive, transformer-based models highlights a relentless drive for innovation. While the inherent ambiguity of language remains the central challenge, the tools we have to address it are more powerful than ever. As we push toward the ultimate goal of true natural language understanding, we not only create more intuitive and powerful applications but also unlock deeper insights into the nature of language, thought, and intelligence itself.

1.7 References

Allen, J. (1995). *Natural language understanding* (2nd ed.). Benjamin/Cummings Publishing.

Anthropic. (2024). *The Claude 3 model family: Opus, Sonnet, Haiku.* https://www.anthropic.com/news/claude-3-family

Bates, M. (1995). Models of natural language understanding. *Proceedings of the National Academy of Sciences, 92*(22), 9977–9982. https://doi.org/10.1073/pnas.92.22.9977

Bender, E. M., Gebru, T., McMillan-Major, A., & Shmitchell, S. (2021). On the dangers of stochastic parrots: Can language models be too big? *Proceedings of the 2021 ACM Conference on Fairness, Accountability, and Transparency*, 610–623. https://doi.org/10.1145/3442188.3445922

Berstler, S. (2024). The Grice is right: Grice's non-cooperation problem and the structure of conversation. *Philosophical Perspectives, 38*(1), 26–40. https://doi.org/10.1111/phpe.12199

Bolukbasi, T., Chang, K. W., Zou, J. Y., Saligrama, V., & Kalai, A. T. (2016). Man is to computer programmer as woman is to homemaker? Debiasing word embeddings. *Advances in Neural Information Processing Systems, 29*, 4356–4364.

Brown, T. B., Mann, B., Ryder, N., Subbiah, M., Kaplan, J., Dhariwal, P., Neelakantan, A., Shyam, P., Sastry, G., Askell, A., Agarwal, S., Herbert-Voss, A., Krueger, G., Henighan, T., Child, R., Ramesh, A., Ziegler, D. M., Wu, J., Winter, C., ... Amodei, D. (2020). Language models are few-shot learners. *Advances in Neural Information Processing Systems, 33*, 1877–1901.

Chomsky, N. (1957). *Syntactic structures.* Mouton de Gruyter.

Cohen, P. R., & Perrault, C. R. (1979). Elements of a plan-based theory of speech acts. *Cognitive Science, 3*(3), 177-212. https://doi.org/10.1207/s15516709cog0303_1

Dale, R. (2021). NLP in the real world: A survey of the application of natural language processing. *Natural Language Engineering, 27*(5), 521–571. https://doi.org/10.1017/S135132492100018X

Devlin, J., Chang, M. W., Lee, K., & Toutanova, K. (2019). BERT: Pre-training of deep bidirectional transformers for language understanding. *Proceedings of NAACL-HLT 2019*, 4171–4186. https://doi.org/10.18653/v1/N19-1423

Dikken, M. den. (2019). Norbert Hornstein, Howard Lasnik, Pritty Patel-Grosz and Charles Yang (eds.): Syntactic structures after 60 years. The impact of the Chomskyan revolution in linguistics. *Acta Linguistica, 66*(3), 429–444. https://doi.org/10.1556/2062.2019.66.3.6

Dong, X. L., Moon, S., Xu, Y. E., Malik, K., & Yu, Z. (2023). Towards next-generation intelligent assistants leveraging LLM techniques. *Proceedings of the 29th ACM SIGKDD Conference on Knowledge Discovery and Data Mining*, 5792–5793. https://doi.org/10.1145/3580305.3599572

Goldberg, Y. (2017). *Neural network methods for natural language processing.* Morgan & Claypool Publishers.

Google. (2023). *Gemini: A family of highly capable multimodal models.* arXiv preprint. https://doi.org/10.48550/arXiv.2312.11805

Grice, H. P. (1975). Logic and conversation. In P. Cole & J. L. Morgan (Eds.), *Syntax and semantics*, Vol. 3: Speech acts (pp. 41–58). Academic Press.

Hirschberg, J., & Manning, C. D. (2015). Advances in natural language processing. *Science, 349*(6245), 261–266. https://doi.org/10.1126/science.aaa8685

Hu, Y. (2024). A socio-cognitive reinterpretation of Grice's theory of conversation. *Intercultural Pragmatics, 21*(1), 99–119. https://doi.org/10.1515/ip-2024-0004

Hutchins, W. J. (1995). The whisky was invisible, or persistent myths of MT. *MT News International, 11*, 17–18.

Jurafsky, D., & Martin, J. H. (2023). *Speech and language processing* (3rd ed.). Prentice Hall.

Kim, J.-W., Yoon, H., & Jung, H.-Y. (2022). Improved spoken language representation for intent understanding in a task-oriented dialogue system. *Sensors (Basel, Switzerland), 22*(4), 1509. https://doi.org/10.3390/s22041509

Manning, C. D., & Schütze, H. (1999). *Foundations of statistical natural language processing.* MIT Press.

McShane, M. (2017). Natural language understanding (NLU, not NLP) in cognitive systems. *The AI Magazine, 38*(4), 43–56. https://doi.org/10.1609/aimag.v38i4.2745

Min, B., Ross, H., Sulem, E., Veyseh, A. P. B., Nguyen, T. H., Sainz, O., Agirre, E., Heintz, I., & Roth, D. (2024). Recent advances in natural language processing via large pre-trained language models: A survey. *ACM Computing Surveys, 56*(2), Article 30. https://doi.org/10.1145/3605943

Nadeau, D., & Sekine, S. (2007). A survey of named entity recognition and classification. *Lingvisticae Investigationes, 30*(1), 3–26. https://doi.org/10.1075/li.30.1.02nad

Naik, D., Naik, I., & Naik, N. (2024). Large data begets large data: Studying large language models (LLMs) and its history, types, working, benefits and limitations. In S. Prajapat, P. Jenkins, N. Naik, & P. Grace (Eds.), *Contributions Presented at The International Conference on Computing, Communication, Cybersecurity and AI, July 3–4, 2024, London, UK* (Vol. 884, pp. 293–314). Springer. https://doi.org/10.1007/978-3-031-74443-3_18

Navigli, R. (2009). Word sense disambiguation: A survey. *ACM Computing Surveys, 41*(2), 1–69. https://doi.org/10.1145/1459352.1459355

Ohmer, X., Bruni, E., & Hupke, D. (2024). From form(s) to meaning: Probing the semantic depths of language models using multisense consistency. *Computational Linguistics - Association for Computational Linguistics, 50*(4), 1507–1556. https://doi.org/10.1162/coli_a_00529

Rabiner, L. R., & Juang, B. H. (1993). *Fundamentals of speech recognition.* Prentice-Hall.

Shadbolt, N., Berners-Lee, T., & Hall, W. (2006). The semantic web revisited. *IEEE Intelligent Systems, 21*(3), 96–101. https://doi.org/10.1109/MIS.2006.62

Sutskever, I., Vinyals, O., & Le, Q. V. (2014). Sequence to sequence learning with neural networks. *Advances in Neural Information Processing Systems, 27*, 3104–3112.

Touvron, H., Martin, L., Stone, K., Albert, P., Almahairi, A., Babaei, Y., Bashlykov, N., Batra, S., Bhargava, P., Bhosale, S., Bikel, D., Blecher, L., Ferrer, C. C., Chen, M., Cucurull, G., Esiobu, D., Fernandes, J., Fu, J., Fu, W., ... Scialom, T. (2023). *Llama 2: Open foundation and fine-tuned chat models.* arXiv preprint. https://doi.org/10.48550/arXiv.2307.09288

Vaswani, A., Shazeer, N., Parmar, N., Uszkoreit, J., Jones, L., Gomez, A. N., Kaiser, Ł., & Polosukhin, I. (2017). Attention is all you need. *Advances in Neural Information Processing Systems, 30*, 5998–6008.

Winograd, T. (1972). *Understanding natural language.* Academic Press.

Yarowsky, D. (1995). Unsupervised word sense disambiguation rivaling supervised methods. *Proceedings of the 33rd Annual Meeting on Association for Computational Linguistics*, 189–196. https://doi.org/10.3115/981658.981684

1.8 Glossary

Ambiguity: The property of language where a word, phrase, or sentence can have multiple possible meanings, creating a primary challenge for computational interpretation.

Artificial Intelligence (AI): A broad field focused on creating intelligent agents, which are systems that can reason, learn, and act autonomously. NLP is a subfield of AI.

Attention Mechanism: A technique used in neural networks, particularly transformers, that allows a model to dynamically weigh the importance of different parts of an input sequence when processing information.

Computational Linguistics: An interdisciplinary field that uses computational techniques to study and model human language. It provides the theoretical foundations for NLP.

Deep Learning: A subfield of machine learning that uses artificial neural networks with multiple layers to analyze complex patterns in data.

Large Language Models (LLMs): Deep learning models with a massive number of parameters, trained on vast amounts of text data, capable of generating coherent and contextually relevant human-like text.

Natural Language Processing (NLP): A field of computer science, linguistics, and artificial intelligence focused on enabling computers to understand, interpret, and generate human language.

Natural Language Understanding (NLU): A subfield of NLP focused on achieving a deeper, human-like comprehension of language, including intent, context, and meaning.

Preprocessing: The crucial first step in an NLP pipeline that involves cleaning and structuring raw text into a format a machine can understand, using techniques like tokenization and lemmatization.

Transformer: A deep learning architecture based on the attention mechanism, which has become the standard for modern NLP models.

1.9 Review Questions and Discussion Topics

1. **Ambiguity:** Using the sentence, "The old man's glasses were on the table," provide a concrete example of both lexical and syntactic ambiguity that a

machine might encounter.

2. **Evolution of NLP:** Describe the evolution of techniques in NLP by classifying them into the three main paradigms discussed in the chapter.

3. **NLU vs. NLP:** Explain the relationship between NLP and NLU. Is it possible to have NLP without NLU? Is it possible to have NLU without NLP?

4. **The Transformer:** What is the transformer architecture, and what key mechanism is responsible for its power? Briefly explain how this mechanism works.

5. **Societal Impact:** The development of LLMs has been compared to the invention of the printing press. Discuss the validity of this analogy. What are the most significant positive and negative societal consequences you foresee?

6. **Connecting Concepts:** How do modern large language models (LLMs), with their transformer architecture, attempt to solve the fundamental "Challenge of Ambiguity" (lexical, syntactic, and semantic) described in Section 1.2.1? In what ways do they succeed, and where do they still fall short?

Chapter 2

Natural Language Processing and Artificial Intelligence

2.1 Introduction

From translating languages in real-time on our smartphones to summarizing complex legal documents in seconds, the interaction between humans and computers has never been more natural. This quiet revolution is powered by natural language processing (NLP), a critical branch of artificial intelligence (AI) that is reshaping our world. AI, in its broadest sense, is the scientific pursuit of imbuing machines with cognitive faculties analogous to human intellect. Within this vast domain is the field of machine learning (ML), which focuses on algorithms that learn patterns from data, and within that is deep learning (DL), which uses complex neural networks to do so. NLP sits at the intersection of these fields, leveraging their power to focus on a singular, profound challenge: enabling computers to interpret, manipulate, and ultimately understand human language (Jurafsky & Martin, 2023).

This book is dedicated to the specific principles and architectures of NLP. As such, it presumes the reader has a foundational understanding of machine learning, including the distinction between supervised and unsupervised learning, the role of training data, and the basic mechanics of gradient descent. While we will explain how these concepts are applied to language, we will not derive their mathematical foundations from first principles. For readers seeking a comprehensive treatment of the broader field of AI or the mathematical underpinnings of deep learning, we highly recommend foundational texts such as *Artificial Intelligence: A Modern Approach* by Russell and Norvig (2021) and *Deep Learning* by Goodfellow, Bengio, and Courville (2016).

Our focus here is on the symbiotic relationship between AI and NLP. The significance of NLP is immense; it promises to automate complex language-based tasks, extract critical insights from unstructured text, and generate coherent content, thereby impacting countless industries. This progress is deeply intertwined with the learning paradigms of ML and DL, which allow systems to master intricate linguistic patterns from vast datasets. Consequently, NLP now serves as the crucial linguistic interface for a myriad of AI applications, making sophisticated technology more intuitive and accessible. This chapter explores this synergy, detailing the evolution of core technologies as they have been adapted for language, surveying key applications, and discussing the persistent challenges that define the future of truly intelligent systems.

2.2 Artificial Intelligence: The Foundation

Artificial intelligence represents the broad discipline focused on creating systems capable of performing tasks that have traditionally required human intelligence. This includes the ability to perceive the environment, reason logically, learn from experience, solve complex problems, and adapt to new situations (Luger, 2009; Bochkova, 2025). The central goal is the development of intelligent systems adept at managing complex tasks through these advanced capabilities, a pursuit that traces its modern origins to the Dartmouth Workshop in 1956 (McCarthy et al., 1955).

The influence of AI is already pervasive. In healthcare, it accelerates drug discovery and aids in interpreting medical images (Esteva et al., 2017; Deckker & Sumanasekara, 2025). Manufacturing leverages AI for predictive maintenance to preempt equipment failure, while the financial sector employs it for sophisticated fraud prevention. We interact with AI daily through search engines, voice-activated assistants, and chatbots. When combined with machine learning, AI vastly enhances our ability to analyze data, enabling faster decisions and more effective predictive analytics (Bishop, 2006; Bhamare & Suryawanshi, 2018). This broad applicability underscores AI's power to solve problems and enhance productivity, with NLP playing a vital role as a translator, enabling these systems to communicate naturally with us.

2.3 Natural Language Processing: AI's Linguistic Branch

Natural language processing (NLP) is the specialized field of AI dedicated to equipping machines to read, comprehend, generate, and respond to human language—whether spoken or written—in a manner that mirrors our own communication (Haralambous, 2024; Tsujii, 2021). It exists at a vibrant intersection of computer science, AI, and linguistics, with the latter providing the essential theoretical understanding of language structure, semantics, and pragmatics needed for robust model design (Chomsky, 1957). The core task of an NLP system is to process unstructured text or speech, converting its inherent complexities and ambiguities into a structured format that a computer can analyze to extract information and discern meaning. The ultimate ambition extends beyond simple pattern recognition to a genuine understanding of linguistic subtleties like intent, emotion, and context (Allen, 1995; Karanikolas et al., 2023). This capacity is fundamental for any AI system designed for effective human engagement, from smartphone voice assistants and email spam filters to automated language translation tools.

As illustrated in Figure 2.1, NLP is a subfield of AI, and it heavily leverages techniques from machine learning (ML) and deep learning (DL), which are themselves subsets of AI (Goodfellow et al., 2016).

Figure 2.1

The Hierarchy of Modern AI

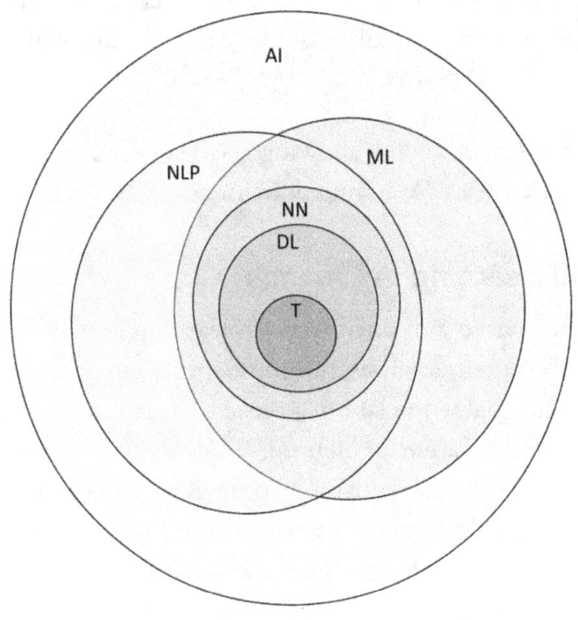

AI: Artificial Intelligence
ML: Machine Learning
DL: Deep Learning
NN: Neural Network
NLP: Natural Language Processing
T: Transformer

2.4 How AI Techniques Power NLP: An Evolutionary Journey

2.4.1 Machine Learning (ML): The Algorithmic Engine

Machine learning, a core subset of AI, provides the algorithmic foundation for modern NLP. It enables machines to learn from data and improve their performance without being explicitly programmed for every scenario (Mitchell, 1997; Shankar, 2022). ML algorithms are indispensable for grappling with the complexities of human language, empowering systems to perform critical tasks like text classification (e.g., spam filtering), sentiment analysis (discerning emotional tone), and named entity recognition (identifying people, places, and organizations in text) (Aggarwal & Zhai, 2012; Minaee et al., 2021).

Two primary paradigms are applied in this context (Murphy, 2012):

- **Supervised Learning** trains models on labeled datasets, where human-provided annotations (such as part-of-speech tags or sentiment labels) guide the model's interpretation. Common algorithms include Support Vector Machines (SVMs) and Bayesian networks.

- **Unsupervised Learning** operates on unlabeled text, discovering hidden structures and relationships without predefined guidance. Techniques like document clustering fall into this category.

These statistical learning methods are fundamental, allowing NLP systems to process language data and extract meaning effectively (Manning & Schütze, 1999).

Supervised vs. Unsupervised Learning in Practice

To make the distinction between supervised and unsupervised learning concrete, consider two common NLP tasks: sentiment analysis and topic modeling. In a supervised sentiment analysis task, the goal is to classify a piece of text into predefined categories such as 'positive,' 'negative' or 'neutral.' To achieve this, the model requires a large, labeled dataset. A data scientist would need to collect, for example, 100,000 movie reviews and have human annotators meticulously label each one. The model is then trained on these examples, learning to associate certain words and phrases (e.g., 'brilliant,' 'amazing') with the 'positive' label and others (e.g., 'terrible,' 'dull') with the 'negative' label. The model's success depends entirely on the quality and quantity of the human-provided supervision.

In stark contrast, an **unsupervised topic modeling** task operates without any human-provided labels. Imagine you have a collection of 50,000 unlabeled news articles from the past month. The goal is to discover the main themes or topics present in this collection. Using an unsupervised algorithm such as latent Dirichlet allocation (LDA), the model analyzes the statistical co-occurrence of words across all documents. It might discover that words like 'election,' 'vote,' 'campaign,' and 'poll' frequently appear together and group them into a latent ('hidden' or not directly observable) *politics* topic. Simultaneously, it might find that words like 'market,' 'stock,' 'trade,' and 'economy' co-occur, forming a *finance* topic. The algorithm discovers these themes organically from the data's inherent structure, without prior knowledge of the categories. This makes unsupervised learning an incredibly powerful tool for exploratory data analysis on large, unstructured text corpora.

2.4.2 From Text to Numbers: The Challenge of Representation

A core challenge in NLP is that computers do not understand words; they understand numbers. Therefore, text must be converted into a numerical format—a process called vectorization—before any learning can occur (Turney & Pantel, 2010; Garg et al., 2018). Early methods included:

- **Bag-of-Words (BoW):** This model represents text as a simple collection of words, disregarding grammar and word order while tracking word frequencies. It creates a vector space where each dimension corresponds to a word in the vocabulary (Harris, 1954; Singh et al., 2024).

- **Term Frequency-Inverse Document Frequency (TF-IDF):** A more sophisticated statistical measure that evaluates how important a word is to a specific document within a larger collection. It increases the weight of terms that are frequent in a document but rare across the corpus (Jones, 1972; Jiang et al., 2021).

While useful, these methods create sparse representations that fail to capture the semantic relationships between words. The words 'king' and 'queen' are treated as completely separate, unrelated entities, revealing a critical need for a more nuanced approach (Manning & Schütze, 1999).

2.4.3 Neural Networks and Word Embeddings: Capturing Meaning

The advent of neural networks, computational models inspired by the structure of the human brain, marked a turning point for NLP (Rumelhart et al., 1986). These networks consist of interconnected layers of nodes that process information, allowing them to learn complex patterns from data. A crucial application of neural networks in NLP is the creation of **word embeddings**—dense vector representations of words. Unlike sparse vectors from BoW or TF-IDF, embeddings capture semantic relationships, positioning words with similar meanings closer together in a multi-dimensional space. Techniques like Word2Vec (Mikolov et al., 2013; Johnson et al., 2024) and GloVe (Pennington et al., 2014; Kulkarni et al., 2020) generate these powerful embeddings, providing models with a much richer understanding of word meaning and context, which significantly boosts performance on all subsequent NLP tasks.

Figure 2.2

The Evolution of Text Representation

Classic Approach: Sparse Vectors

(e.g., Bag-of-Words)

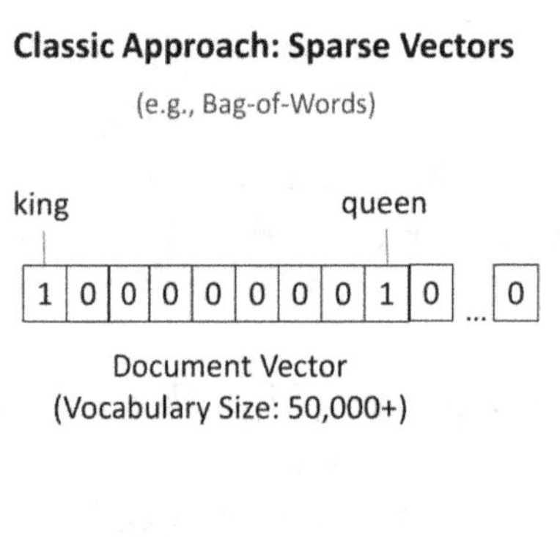

Document Vector
(Vocabulary Size: 50,000+)

Modern Approach: Dense Embeddings

(e.g., Word2Vec / BERT).

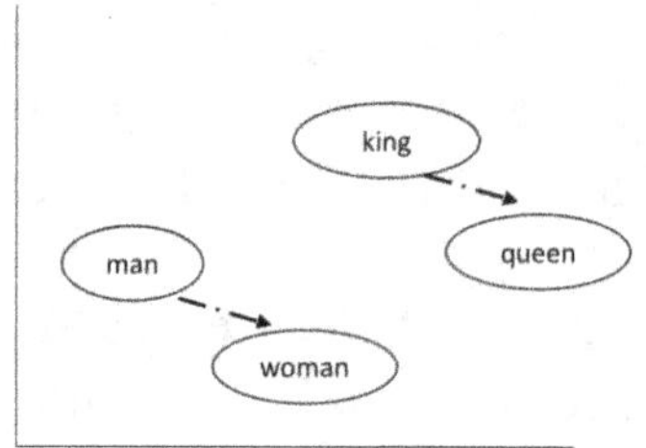

Limitation: Words are treated as independent, unrelated items. The model has no concept that "king" and "queen" are similar.

Advantage: Semantic relationships are captured as distances. The model learns that "king" is to "queen" as "man" is to "woman".

The Magic of Embeddings: Semantic Arithmetic

The power of word embeddings stems from the **distributional hypothesis**, a core linguistic theory that posits that words that appear in similar contexts tend to have similar meanings. Word embedding algorithms like Word2Vec operationalize this idea by learning a vector for each word based on its neighbors in a large text corpus. The resulting high-dimensional vector space is not random; it is structured to capture complex semantic relationships. The most famous and intuitive demonstration of this is the concept of **semantic arithmetic**. In a well-trained embedding space, vector operations can reveal linguistic analogies. The canonical example is that if you take the vector for the word 'King', subtract the vector for 'Man', and add the vector for 'Woman', the resulting vector will be closer to the vector for 'Queen' than to any other word in the vocabulary. This can be expressed as: vector('King') - vector('Man') + vector('Woman') ≈ vector('Queen'). This astonishing result shows that the model has not just memorized words, but has learned abstract relationships like gender and royalty and encoded them into the geometric structure of the vector space. This ability to represent word meaning in a way that allows for such analogical reasoning was a revolutionary leap beyond simple frequency counts (Mikolov et al., 2013).

2.4.4 Deep Learning Architectures: A Story of Evolution

Deep learning (DL), which utilizes deep neural networks with many layers, has driven transformative breakthroughs in NLP by automatically learning complex, hierarchical representations directly from language data (LeCun et al., 2015). This has led to the evolution of specialized architectures for processing language with increasing sophistication.

- **Recurrent Neural Networks (RNNs): The Dawn of Sequence Processing.** Because language is sequential, RNNs were a natural fit. They utilize a hidden state, or a form of memory, to retain information about previous elements in a sequence when processing the current one. However, they struggled to maintain information over long sequences, a problem known as the vanishing gradient (Bengio et al., 1994; Mienye et al., 2024).

- **LSTMs and GRUs: Overcoming Amnesia.** To address the shortcomings of RNNs, advanced variants were developed. Long short-term memory (LSTM) networks (Hochreiter & Schmidhuber, 1997; Yu et al., 2020) and gated recurrent units (GRUs) (Cho et al., 2014) incorporate sophisticated *gating mechanisms* that carefully control the flow of information, enabling them to effectively learn and remember long-range dependencies in text.

- **The Transformer: A Paradigm Shift with Attention.** The introduction of the transformer architecture in 2017 represented a major paradigm shift (Vaswani et al., 2017). Instead of processing text sequentially like an RNN, transformers use a self-attention mechanism. This allows the model to simultaneously weigh the importance of all other words in the input sequence, regardless of their distance, enabling a more powerful, holistic understanding of context. This design also enables parallel processing, drastically reducing training time. This architecture now forms the foundation of nearly all state-of-the-art models, including BERT and the GPT series (Devlin et al., 2019; Brown et al., 2020), as well as more recent large-scale models like Meta's Llama series (Touvron et al., 2023) and Google's Gemini (Gemini Team, 2023).

The evolution from RNNs to transformers represents a fundamental shift in architectural philosophy. RNNs and LSTMs are inherently **sequential**, processing a sentence one token at a time from left to right. This approach is intuitive, mirroring how humans read, but it imposes a severe computational bottleneck. The calculation for the tenth word cannot begin until the calculation for the ninth word is complete, making it impossible to fully parallelize the process across the sequence. The transformer, by contrast, adopts a **parallel processing** paradigm. Its self-attention mechanism allows it to compute representations for every word in the sentence simultaneously, creating a direct path between every pair of words. This ability to process the entire sequence at once, rather than one word at a time, is the key innovation that unlocked massive scalability. It allowed researchers to train much larger and deeper models on vastly more data than was ever feasible with sequential architectures, directly leading to the modern era of large language models (Vaswani et al., 2017).

2.5 Challenges and Research Directions

Despite tremendous progress, both AI and NLP confront substantial hurdles that shape their research agendas.

2.5.1 Overarching AI Challenges

The broader field of AI grapples with significant ethical, legal, and technical issues.

- **Bias and Fairness:** A primary concern is algorithm bias, where systems perpetuate or amplify societal biases present in their training data, leading to unfair outcomes. For instance, word embeddings have been shown to encode gender stereotypes (Bolukbasi et al., 2016; Gallegos et al., 2023). Ensuring fairness and accountability is a critical area of research (O'Neil, 2016).

- **Data and Privacy:** The massive data requirements of modern AI models raise significant privacy and security concerns. These models can sometimes inadvertently 'memorize' and later reveal sensitive information from their training data. This has necessitated the development of techniques like **differential privacy**, a formal method that provides a strong, mathematical guarantee of privacy. It works by injecting a carefully calibrated amount of statistical 'noise' into the data or the results of an analysis. This noise is just large enough to make it impossible to determine whether any single individual's data was included in the training set, thus protecting their privacy, while being small enough not to significantly degrade the model's overall accuracy (Dwork, 2008; Zhu et al., 2017; Sørensen, 2023).

- **Explainability (XAI):** The "black box" nature of many deep learning models hinders our ability to understand their decision-making processes, a major barrier to trust in critical applications such as medicine and finance. This has spurred the development of XAI techniques like LIME (Ribeiro et al., 2016; Danilevsky et al., 2020).

- **Robustness and Safety:** Preventing models from generating 'hallucinations' (plausible but factually incorrect outputs) and ensuring they operate safely and reliably remain key technical challenges. This has spurred the development of mitigation techniques and a greater focus on responsible AI practices (Bender et al., 2021).

The "Black Box" Problem and the Need for XAI

The need for explainable AI (XAI) is not merely an academic concern; it is a critical requirement for the responsible deployment of AI in high-stakes environments. Consider an NLP model designed to assist a bank in processing loan applications by analyzing applicants' written personal statements. If the model flags an application for denial, the "black box" problem becomes immediately apparent. The loan officer and, more importantly, the applicant have a right to know *why* the decision was made. Was the application denied because the applicant's statement contained keywords associated with high financial risk? Or did the model learn a spurious, biased correlation from its training data, unfairly penalizing the applicant for using a particular dialect or a non-standard sentence structure associated with a specific demographic group? Without the ability to interrogate the model's reasoning, it is impossible to audit it for fairness, debug its errors, or ensure it complies with financial regulations that prohibit discrimination. This opacity is a fundamental barrier to trust and accountability. The goal of XAI is to develop methods that can shed light on this black box, providing clear, human-understandable justifications for a model's decisions, thereby enabling the kind of scrutiny essential to any system with a significant impact on people's lives (Gerlings et al., 2022; Nagahisarchoghaei et al., 2023).

2.5.2 Specific NLP Challenges

NLP inherits the challenges of AI and adds difficulties intrinsic to the nature of human language.

- **Ambiguity:** Language is fundamentally ambiguous. Words can have multiple meanings (lexical), sentence structures can be interpreted in multiple ways (syntactic), and true meaning is often heavily dependent on context (semantic). (Manning & Schütze, 1999).

- **Diversity and Data Scarcity:** While NLP has excelled in high-resource languages like English, thousands of low-resource languages lack the large datasets needed to train effective models. Handling dialects, slang, and cultural references remains a major hurdle (Joshi et al., 2020). For a deeper look at the advanced techniques used to address these challenges, see Appendix I.

- **Ethical Bias in Language:** Models trained on vast internet corpora inevitably absorb societal biases related to gender, race, and other attributes, which can then surface in applications for hiring or content moderation (Buolamwini & Gebru, 2018; Ankita et al., 2024).

- **Commonsense Reasoning:** Achieving a deeper semantic understanding—moving beyond pattern matching to grasp intent and apply commonsense knowledge—represents a long-term frontier for the field. This challenge requires models to have an implicit understanding of how the world works, a problem that has been a goal of AI from its inception (McCarthy, 1960; Chanin & Hunter, 2023).

2.6 Conclusion

Natural language processing (NLP) and artificial intelligence (AI) are deeply symbiotic. Advances in AI, particularly deep learning, provide the computational engine for NLP to master language, while NLP serves as the essential interface that makes powerful AI systems accessible and interactive. However, this interdependence means NLP inherits AI's core challenges—such as algorithmic bias, data privacy, and the "black box" nature of models—while also confronting its own difficulties rooted in linguistic ambiguity and data scarcity for most of the world's languages.

To overcome these hurdles, the research agendas for both fields are converging on a shared vision for models that are not only accurate but also robust, fair, and transparent. The path forward involves moving beyond purely data-driven methods to incorporate world knowledge, commonsense reasoning, and techniques such as few-shot learning—the ability of a model to learn a new task from a handful of examples rather than thousands—to address data limitations. The future potential of AI and NLP hinges on this synergy, prioritizing the development of robust ethical frameworks and explainable systems (XAI) to responsibly unlock new insights into language and intelligence.

2.7 References

Aggarwal, C. C., & Zhai, C. (2012). A survey of text classification algorithms. In C. C. Aggarwal & C. Zhai (Eds.), *Mining text data* (pp. 163–222). Springer. https://doi.org/10.1007/978-1-4614-3223-4_6

Allen, J. (1995). *Natural language understanding* (2nd ed.). Benjamin/Cummings Publishing.

Ankita B., ChienChen, H., Lauren, P., & Lydia, O. (2024). Impact of explainable AI on reduction of algorithm bias in facial recognition technologies. *Systems and Information Engineering Design, SIEDS, IEEE Symposium*, 85–89. https://doi.org/10.1109/SIEDS61124.2024.10534745

Bender, E. M., Gebru, T., McMillan-Major, A., & Shmitchell, S. (2021). On the dangers of stochastic parrots: Can language models be too big? *In Proceedings of the 2021 ACM Conference on Fairness, Accountability, and Transparency* (pp. 610–623). ACM. https://doi.org/10.1145/3442188.3445922

Bengio, Y., Simard, P., & Frasconi, P. (1994). Learning long-term dependencies with gradient descent is difficult. *IEEE Transactions on Neural Networks, 5*(2), 157–166. https://doi.org/10.1109/72.279181

Bhamare, D., & Suryawanshi, P. (2018). Review on reliable pattern recognition with machine learning techniques. *Fuzzy Information and Engineering, 10*(3), 362–377. https://doi.org/10.1080/16168658.2019.1611030

Bishop, C. M. (2006). *Pattern recognition and machine learning.* Springer.

Bochkova, A. A. (2025). Artificial intelligence: Strategies and methods for solving complex problems. *Dependability, 25*(1), 46–57. https://doi.org/10.21683/1729-26462025-25-1-46-57

Bolukbasi, T., Chang, K. W., Zou, J. Y., Saligrama, V., & Kalai, A. T. (2016). Man is to computer programmer as woman is to homemaker? Debiasing word embeddings. *Advances in Neural Information Processing Systems, 29.*

Brown, T. B., Mann, B., Ryder, N., Subbiah, M., Kaplan, J., Dhariwal, P., ... & Amodei, D. (2020). Language models are few-shot learners. *Advances in Neural Information Processing Systems, 33,* 1877–1901.

Buolamwini, J., & Gebru, T. (2018). Gender shades: Intersectional accuracy disparities in commercial gender classification. *Proceedings of Machine Learning Research, 81,* 77–91.

Chanin, D., & Hunter, A. (2023). *Neuro-symbolic commonsense social reasoning.* arXiv preprint arXiv:2303.08264. https://doi.org/10.48550/arXiv.2303.08264

Cho, K., Van Merriënboer, B., Gulcehre, C., Bahdanau, D., Bougares, F., Schwenk, H., & Bengio, Y. (2014). Learning phrase representations using RNN encoder-decoder for statistical machine translation. *In Proceedings of the 2014 Conference on Empirical Methods in Natural Language Processing (EMNLP)* (pp. 1724–1734). https://doi.org/10.3115/v1/D14-1179

Chomsky, N. (1957). *Syntactic structures.* Mouton de Gruyter.

Danilevsky, M., Qian, K., Aharonov, R., Katsis, Y., Kawas, B., & Sen, P. (2020). *A survey of the state of explainable AI for natural language processing.* arXiv preprint arXiv:2010.00711. https://doi.org/10.48550/arXiv.2010.00711

Deckker, D., & Sumanasekara, S. (2025). From stethoscopes to supercomputers: The AI revolution in medicine: A review. *World Journal of Advanced Research and Reviews, 26*(1), 1114–1131. https://doi.org/10.30574/wjarr.2025.26.1.1138

Devlin, J., Chang, M. W., Lee, K., & Toutanova, K. (2019). BERT: Pre-training of deep bidirectional transformers for language understanding. *In Proceedings of the 2019 Conference of the North American Chapter of the Association for Computational Linguistics: Human Language Technologies,* Volume 1 (Long and Short Papers) (pp. 4171–4186). https://doi.org/10.18653/v1/N19-1423

Dwork, C. (2008). Differential privacy: A survey of results. *In International Conference on Theory and Applications of Models of Computation* (pp. 1-19). Springer. https://doi.org/10.1007/978-3-540-79228-4_1

Esteva, A., Kuprel, B., Novoa, R. A., Ko, J., Swetter, S. M., Blau, H. M., & Thrun, S. (2017). Dermatologist-level classification of skin cancer with deep neural networks. *Nature, 542*(7639), 115–118. https://doi.org/10.1038/nature21056

Gallegos, J. O., et al. (2023). *Bias and fairness in large language models: A survey.* arXiv preprint arXiv:2309.00770.

Garg, N., Schiebinger, L., Jurafsky, D., & Zou, J. (2018). Word embeddings quantify 100 years of gender and ethnic stereotypes. *Proceedings of the National Academy of Sciences, 115*(16), E3635–E3644. https://doi.org/10.1073/pnas.1720347115

Gemini Team. (2023). *Gemini: A family of highly capable multimodal models.* arXiv preprint arXiv:2312.11805.

Gerlings, J., Jensen, M. S., & Shollo, A. (2022). *Explainable AI, but explainable to whom?* arXiv.Org. https://doi.org/10.48550/arxiv.2106.05568

Goodfellow, I., Bengio, Y., & Courville, A. (2016). *Deep learning.* MIT Press.

Goldberg, Y. (2017). *Neural network methods for natural language processing*. The MIT Press.

Haralambous, Y. (2024). *A course in natural language processing* (1st ed.). Springer International Publishing. https://doi.org/10.1007/978-3-031-27226-4

Harris, Z. S. (1954). Distributional structure. *Word, 10*(2-3), 146–162. https://doi.org/10.1080/00437956.1954.11659520

Hochreiter, S., & Schmidhuber, J. (1997). Long short-term memory. *Neural Computation, 9*(8), 1735–1780. https://doi.org/10.1162/neco.1997.9.8.1735

Jiang, Z., Gao, B., He, Y., Han, Y., Doyle, P., & Zhu, Q. (2021). Text classification using novel term weighting scheme-based improved TF-IDF for internet media reports. *Mathematical Problems in Engineering, 2021*, 1–30. https://doi.org/10.1155/2021/6619088

Johnson, S. J., Murty, M. R., & Navakanth, I. (2024). A detailed review on word embedding techniques with emphasis on word2vec. *Multimedia Tools and Applications, 83*(13), 37979–38007. https://doi.org/10.1007/s11042-023-17007-z

Jones, K. S. (1972). A statistical interpretation of term specificity and its application in retrieval. *Journal of Documentation, 28*(1), 11-21. https://doi.org/10.1108/eb026526

Joshi, P., Santy, S., Budhiraja, A., Bali, K., & Choudhury, M. (2020). The state and fate of linguistic diversity and inclusion in the NLP world. *In Proceedings of the 58th Annual Meeting of the Association for Computational Linguistics* (pp. 6282–6293). https://doi.org/10.18653/v1/2020.acl-main.560

Jurafsky, D., & Martin, J. H. (2023). *Speech and language processing* (3rd ed.). Prentice Hall.

Karanikolas, N., Manga, E., Samaridi, N., Tousidou, E., & Vassilakopoulos, M. (2023). Large language models versus natural language understanding and generation. In C. Marinagi, A. Kakarountas, N. N. Karanikolas, I. Voyiatzis, & M. G. Vassilakopoulos (Eds.), *Proceedings of the 27th Pan-Hellenic Conference on Progress in Computing and Informatic*s (pp. 278–290). ACM. https://doi.org/10.1145/3635059.3635104

Kulkarni, S., Katariya, J. K., & Potika, K. (2020). GloVeNoR: GloVe for node representations with second order random walks. *Proceedings of the ... International Conference on Advances in Social Network Analysis and Mining*, 536–543. https://doi.org/10.1109/ASONAM49781.2020.9381347

LeCun, Y., Bengio, Y., & Hinton, G. (2015). Deep learning. *Nature, 521*(7553), 436–444. https://doi.org/10.1038/nature14539

Luger, G. F. (2009). *Artificial intelligence: Structures and strategies for complex problem solving* (6th ed.). Pearson.

Manning, C. D., & Schütze, H. (1999). *Foundations of statistical natural language processing*. MIT Press.

McCarthy, J. (1960). Programs with common sense. *In Proceedings of the Teddington Conference on the Mechanization of Thought Processes* (pp. 77–84). Her Majesty's Stationery Office.

McCarthy, J., Minsky, M. L., Rochester, N., & Shannon, C. E. (1955). A proposal for the Dartmouth summer research project on artificial intelligence. *AI Magazine, 27*(4), 12.

Mienye, I. D., Swart, T. G., & Obaido, G. (2024). Recurrent neural networks: A comprehensive review of architectures, variants, and applications. *Information (Basel), 15*(9), 517. https://doi.org/10.3390/info15090517

Mikolov, T., Chen, K., Corrado, G., & Dean, J. (2013). *Efficient estimation of word representations in vector space*. arXiv preprint arXiv:1301.3781.

Minaee, S., Kalchbrenner, N., Cambria, E., Nikzad, N., Chenaghlu, M., & Gao, J. (2021). Deep learning based text classification: A comprehensive review. *ACM Computing Surveys (CSUR), 54*(3), 1–40. https://doi.org/10.1145/3439726

Mitchell, T. M. (1997). *Machine learning*. McGraw-Hill.

Murphy, K. P. (2012). *Machine learning: A probabilistic perspective*. MIT Press.

Nagahisarchoghaei, M., Nur, N., Cummins, L., Nur, N., Karimi, M. M., Nandanwar, S., Bhattacharyya, S., & Rahimi, S. (2023). An empirical survey on explainable AI technologies: Recent trends, use-cases, and categories from technical and application perspectives. *Electronics (Basel), 12*(5), 1092. https://doi.org/10.3390/electronics12051092

O'Neil, C. (2016). *Weapons of math destruction: How big data increases inequality and threatens democracy*. Crown.

Pennington, J., Socher, R., & Manning, C. D. (2014). GloVe: Global vectors for word representation. *In Proceedings of the 2014 Conference on Empirical Methods in Natural Language Processing (EMNLP)* (pp. 1532–1543). https://doi.org/10.3115/v1/D14-1162

Ribeiro, M. T., Singh, S., & Guestrin, C. (2016). Why should I trust you?: Explaining the predictions of any classifier. *In Proceedings of the 22nd ACM SIGKDD International Conference on Knowledge Discovery and Data Mining* (pp. 1135–1144). https://doi.org/10.1145/2939672.2939778

Rumelhart, D. E., Hinton, G. E., & Williams, R. J. (1986). Learning representations by back-propagating errors. *Nature, 323*(6088), 533–536. https://doi.org/10.1038/323533a0

Russell, S. J., & Norvig, P. (2021). *Artificial intelligence: A modern approach* (4th ed.). Pearson.

Shankar, U. (2022). *Machine and deep learning algorithms and applications* (A. Spanias, Ed.; 1st ed. 2022.). Springer Nature. https://doi.org/10.1007/978-3-031-03758-0

Singh, U. K., Prabhu Shankar, B., Chinnaiyan, R., & Jain, N. (2024). Machine learning-based text categorization with bag of words. In Y. Singh, P. J. S. Gonçalves, P. K. Singh, & A. K. Kar (Eds.), *Proceedings of International Conference on Recent Innovations in Computing* (pp. 577–587). Springer Nature Singapore. https://doi.org/10.1007/978-981-97-2839-8_40

Sørensen, F. B. (2023). *Privacy-preserving data analysis with differential privacy*.

Touvron, H., Martin, L., Stone, K., Albert, P., Almahairi, A., Babaei, Y., ... & Scialom, T. (2023). *Llama 2: Open foundation and fine-tuned chat models*. arXiv preprint arXiv:2307.09288.

Tsujii, J. (2021). Natural language processing and computational linguistics. *Computational Linguistics, 47*(4), 1–21. https://doi.org/10.1162/coli_a_00420

Turney, P. D., & Pantel, P. (2010). From frequency to meaning: Vector space models of semantics. *Journal of Artificial Intelligence Research, 37*, 141–188. https://doi.org/10.1613/jair.2934

Vaswani, A., Shazeer, N., Parmar, N., Uszkoreit, J., Jones, L., Gomez, A. N., ... & Polosukhin, I. (2017). Attention is all you need. *Advances in Neural Information Processing Systems, 30*.

Yu, Y., Si, X., Hu, C., & Zhang, J. (2020). A review of recurrent neural networks: LSTM cells and network architectures. *Neural Computation, 32*(6), 1235–1270.

https://doi.org/10.1162/neco_a_01199

Zhu, T., Li, G., Zhou, W., & Yu, P. S. (2017). Differentially private data publishing and analysis: A survey. *IEEE Transactions on Knowledge and Data Engineering, 29*(8), 1619–1638. https://doi.org/10.1109/TKDE.2017.2697856

2.8 Glossary

Algorithm Bias: A primary concern where AI systems perpetuate or amplify societal biases present in their training data, which can lead to unfair outcomes. These biases can be related to gender, race, and other attributes absorbed from vast internet corpora.

Artificial Intelligence (AI): The scientific and engineering pursuit of creating systems with cognitive faculties similar to human intellect. Its goal is to develop intelligent systems that can perceive the environment, reason logically, learn from experience, solve complex problems, and adapt to new situations.

Bag-of-Words (BoW): An early method of text representation that models a document as a simple collection of its words, ignoring grammar and word order. It creates a sparse representation that fails to capture semantic relationships.

Deep Learning (DL): A field of AI that uses deep neural networks (models with many layers) to automatically learn complex, hierarchical representations directly from data. It has driven transformative breakthroughs in NLP.

Explainability (XAI): A field of research focused on addressing the "black box" nature of many deep learning models. It aims to make a model's decision-making process understandable, which is critical for building trust in applications like medicine and finance.

Few-shot learning: A type of machine learning model that can learn to recognize new objects or concepts from a very small number of examples—sometimes as few as one or just a handful. It's designed to mimic the human ability to learn quickly from limited information.

Gated Recurrent Units (GRUs): An advanced variant of RNNs that uses "gating mechanisms" to control the flow of information. This allows them to effectively learn and remember long-range dependencies in text, overcoming a key weakness of simple RNNs.

Hallucinations: A key technical challenge where AI models generate plausible but factually incorrect outputs.

Long Short-Term Memory (LSTM): A type of advanced RNN that incorporates "gating mechanisms" to manage the flow of information. This design enables the network to effectively learn and remember information over long sequences, addressing a major shortcoming of basic RNNs.

Machine Learning (ML): A core subset of AI that provides the algorithmic foundation for modern NLP. It enables machines to learn from data and improve their performance without being explicitly programmed for every scenario.

Natural Language Processing (NLP): A specialized branch of AI focused on enabling computers to interpret, manipulate, understand, and generate human language, whether spoken or written. It exists at the intersection of computer science, AI, and linguistics.

Neural Networks: Computational models inspired by the structure of the human brain, consisting of interconnected layers of nodes that process information. They are used in NLP to learn complex patterns and create word embeddings.

Recurrent Neural Networks (RNNs): Neural networks designed to process sequential data, making them a natural fit for language. They use a hidden state, or memory, to retain information about previous elements in a sequence. However, they struggle to maintain information over long sequences due to the vanishing gradient problem.

Self-attention mechanism: The core component of the transformer architecture that allows the model to weigh the importance of all other words in an input sequence simultaneously, regardless of their position. This enables a more holistic understanding of context.

Supervised Learning: A machine learning paradigm that trains models on labeled datasets where human-provided annotations guide the model in its interpretation. Common algorithms include Support Vector Machines and Bayesian networks.

Term Frequency-Inverse Document Frequency (TF-IDF): A statistical measure used for text vectorization that evaluates how important a word is to a specific document within a larger collection.

Transformer: A state-of-the-art architecture introduced in 2017 that represented a major paradigm shift in NLP. It uses a self-attention mechanism rather than sequential processing, enabling parallel processing and a deeper understanding of context. It is the foundation for models like BERT and the GPT series.

Unsupervised Learning: A machine learning paradigm that operates on unlabeled text to discover hidden structures and relationships without predefined guidance. Document clustering is an example of this technique.

Vectorization: The process of converting text into a numerical format so that machine learning can occur.

Word Embeddings: Dense vector representations of words created using neural networks. Unlike sparse vectors, embeddings capture semantic relationships, positioning words with similar meanings closer together in a multi-dimensional space.

2.9 Review Questions and Discussion Topics

1. The chapter highlights that a primary concern in AI is "algorithm bias," in which systems perpetuate societal biases embedded in their training data. Discuss the potential real-world consequences of deploying a biased NLP model in critical sectors like finance or for use in hiring applications. What steps could be taken to mitigate these biases?

2. Explain the evolution of text representation in NLP, starting from early methods like Bag-of-Words and TF-IDF to the development of word embeddings from neural networks. Why was the ability of word embeddings to capture semantic relationships between words considered a "turning point for NLP"?

3. Compare the architectural approaches of recurrent neural networks (RNNs) and the transformer. What fundamental problem did RNNs have with processing long sequences, and how did the transformer's self-attention mechanism solve this issue?

4. The text introduces explainability (XAI) as a response to the "black box" nature of many deep learning models. Why is the ability to understand a model's decision-making process a major barrier to trust and adoption in high-stakes fields like medicine and finance?

5. NLP models have excelled in high-resource languages but face significant challenges with the thousands of low-resource languages due to data scarcity. Imagine you are tasked with developing a translation tool for a language with a small digital footprint. What are the primary obstacles you would face, and how does the challenge of achieving commonsense reasoning complicate this task?

Chapter 3

Language Analysis

3.1 Introduction

Language analysis is an essential discipline through which linguists break down
language components and structures to understand their significance and practical
use. For centuries, this was a purely human endeavor. Today, it stands as the critical
bridge between human language and machine understanding, forming the bedrock
of artificial intelligence. The ability of a machine to translate languages, answer
complex questions, or summarize a lengthy report does not stem from a single,
monolithic *understanding*. Rather, it is the result of a multi-layered analytical process,
often conceptualized as a pipeline where each stage adds a richer layer of
interpretation.

This chapter examines the essential techniques of language analysis, exploring how
machines methodically deconstruct language to process and interpret its meaning.
We will dissect the six primary layers of this process, which represent a classic,
idealized model of computational language analysis:

- **Lexical Analysis:** Examining individual words.

- **Morphological Analysis:** Investigating the internal structure of words.

- **Syntax Analysis:** Focusing on sentence structure and grammar.

- **Semantic Analysis:** Exploring literal word and sentence meaning.

- **Pragmatic Analysis:** Considering the influence of context on
  interpretation.

- **Discourse Analysis:** Investigating the organization of text and
  conversation.

Figure 3.1

The Stack of Language Analysis

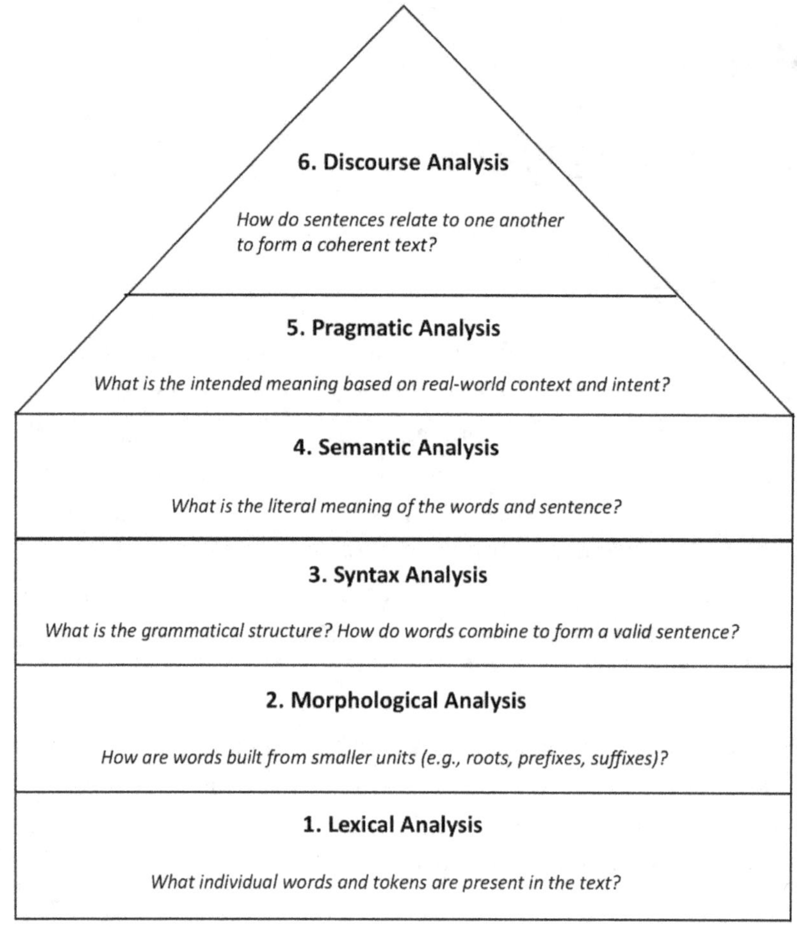

Mastering these analytical methods is vital for anyone in the field of natural language processing (NLP), as they are the foundational techniques that enable computers to process and understand human language (Manning & Schütze, 1999).

3.2 Language Analysis

Language analysis involves systematically breaking down linguistic data to understand its components, structure, meaning, and use. Below, we explore the key types of analysis, demonstrating how each layer builds upon the last to create a holistic understanding, a concept that has been central to NLP since its early development (Winograd, 1972, 1980).

To see these methods in action, we will use a running example. Consider the following short passage:

> "The researcher's glasses were on the bank. She left them there after a long day of observing river otters."

This simple text contains multiple ambiguities that a computer must resolve. Let's see how.

3.2.1 Lexical Analysis

Lexical analysis is the foundational first step in language processing, where a stream of text is broken down into its smallest meaningful units, known as lexemes or tokens. This process, called tokenization, provides the basic building blocks for all higher-level analysis. Each token is then categorized with its part-of-speech (POS) tag—such as noun, verb, or adjective—which describes its grammatical function (Manning & Schütze, 1999). In our example, "The researcher's glasses were on the bank," a lexical analysis would first tokenize the sentence and then tag each word:

- 'The' - Determiner

- 'researcher's' - Possessive Noun

- 'glasses' - Noun

- 'were' - Verb

- 'on' - Preposition

- 'the' - Determiner

- 'bank' - Noun

At this stage, a critical ambiguity emerges. The lexeme 'bank' could refer to a financial institution or the side of a river. Similarly, 'glasses' could refer to eyewear or drinking vessels. Lexical analysis identifies these potential meanings, but it lacks the context to resolve them. Its sole focus is to inventory and categorize the words as they appear. This output is fundamental for applications like search engines and compilers to process raw input (Gusfield, 1997).

3.2.2 Morphological Analysis

While lexical analysis treats words as indivisible units, morphological analysis delves into their internal structure. It examines how words are built from morphemes—the smallest units of meaning, such as roots, prefixes, and suffixes (Haspelmath & Sims, 2010). This deconstruction is vital for understanding the nuances of word meaning and is the basis for spelling checkers and lemmatization tools, which reduce words to their dictionary base form (Jurafsky & Martin, 2023).

Applying this to our example, morphological analysis would break down 'researcher's' into three morphemes:

- 'research' (the root verb)

- '-er' (a suffix that denotes a person who does the action)

- 's' (a possessive suffix)

This analysis reveals that the word describes a person who researches, not just an object. This distinction is crucial for downstream tasks that rely on understanding word relationships (Spencer, 1991).

3.2.3 Syntax Analysis

Syntax analysis, or parsing, takes the tokenized output from lexical analysis and determines the sentence's grammatical structure. It applies a set of grammar rules to understand how words combine to form phrases, clauses, and ultimately, grammatically correct sentences (Chomsky, 1957). While it rigorously enforces grammatical rules, syntax analysis operates without any consideration for the meaning of the words themselves.

For our sentence, "The researcher's glasses were on the bank," syntax analysis would produce a parse tree that identifies:

- Subject: "The researcher's glasses" (a Noun Phrase)

- Verb: 'were' (a Verb)

- Prepositional Phrase: 'on the bank' (which modifies the verb)

The analysis confirms the sentence is grammatically valid. However, it still cannot resolve the ambiguity of 'bank' or 'glasses.' It understands the sentence's architecture but not its meaning. This structural understanding is essential for grammar checkers and machine translation systems (Manning & Schütze, 1999).

Constituency vs. Dependency Parsing

The parse tree described above is the result of **constituency parsing**, which breaks a sentence down into its constituent parts or phrases (e.g., noun phrases, verb phrases). This approach reveals the hierarchical, phrase-based structure of a sentence. However, it is not the only way to model syntax. An alternative and equally powerful paradigm is **dependency parsing**. Instead of focusing on phrases, dependency parsing models the grammatical relationships between individual words. It represents the sentence as a directed graph, where words are nodes and edges are labeled dependencies (e.g., nsubj for nominal subject, dobj for direct object), indicating which words modify or depend on other words, with the main verb typically acting as the root of the tree.

For our example sentence, "The researcher's glasses were on the bank," a dependency parse would reveal the following relationships:

- *'were'* is the root of the sentence.

- *'glasses'* is the nominal subject (nsubj) of *were*.

- *'The'* and *'researcher's'* are modifiers of glasses.

- *'bank'* is the nominal modifier (nmod) of *were*, connected by the preposition *on*.

This dependency view offers several advantages. It is often more effective for languages with free word order, as the direct links between words are not dependent on their sequential position. Furthermore, explicit labeling of grammatical functions (subject, object, etc.) can be a more direct and useful input for downstream semantic tasks, such as information extraction, which are often concerned with these relationships. Both constituency and dependency parsing provide valid and valuable views of syntactic structure, and the choice between them often depends on the specific requirements of the NLP application.

Figure 3.2

Two Approaches to Syntactic Parsing

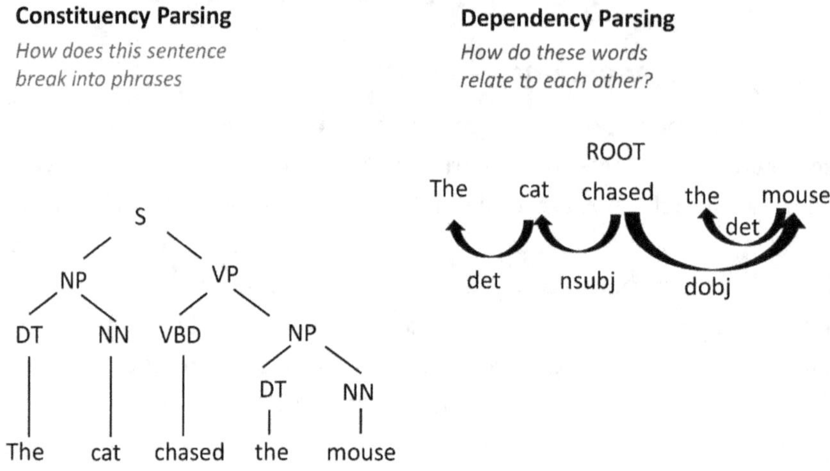

3.2.4 Semantic Analysis

Semantic analysis is where the process of interpretation truly begins. Its goal is to determine the literal meaning of words, phrases, and sentences within their given context. It moves beyond grammar to ask, "What is actually being communicated?" (Jurafsky & Martin, 2023). This stage addresses the ambiguities unresolved by lexical and syntactic analysis, a task often framed as word-sense disambiguation (Navigli, 2009).

To resolve the meaning of 'bank,' semantic analysis would look at the surrounding words and the broader context provided by our second sentence: "She left them there after a long day of observing river otters."

- The phrase "river otters" strongly suggests a context of nature and water. Therefore, the analysis can infer that 'bank' refers to a riverbank rather than a financial institution.

- Similarly, the act of 'observing' makes it far more likely that "glasses" refers to eyewear.

By relating words to each other, semantic analysis disambiguates meaning. This capability is critical for search engines to understand query intent and for virtual assistants to grasp the meaning of commands (Lappin, 1996).

Deeper Semantics: Semantic Role Labeling (SRL)

While word sense disambiguation focuses on individual word meanings, a deeper level of semantic analysis seeks to understand the meaning of the entire event described in a sentence. This is the goal of **semantic role labeling (SRL)**. SRL is the task of identifying the predicate (typically a verb or verb-like word) and labeling the semantic roles of its participants. It aims to answer the fundamental question of "who did what to whom, when, where, and how?" The assigned labels are abstract semantic roles rather than purely syntactic ones. Key roles include:

- **AGENT:** The entity that intentionally performs the action.

- **PATIENT** (or **THEME**): The entity that undergoes the action or is affected by it.

- **INSTRUMENT:** The tool used to perform the action.

- **LOCATION:** Where the action took place.

- **TEMPORAL:** When the action took place.

Let's apply SRL to the second sentence of our running example: "She left them there after a long day of observing river otters." Focusing on 'observing', an SRL system would identify:

- **PREDICATE**: observing

- **AGENT**: She

- **PATIENT**: river otters

This analysis provides a structured, semantic representation of the event that is more abstract than a syntactic parse. It makes the core meaning explicit, which is incredibly valuable for tasks like question answering (to answer "Who observed the otters?"), machine translation, and text summarization.

3.2.5 Pragmatic Analysis

Language often communicates more than what is literally stated. Pragmatic analysis examines this hidden layer of meaning by considering the broader context, the speaker's intent, and the relationship between participants (Levinson, 1983). While semantics concerns literal meaning, pragmatics concerns intended meaning. A classic example is interpreting "Can you pass the salt?" not as a question about physical ability, but as a polite request (Austin & Urmson, 1962). In our running example, the pragmatic analysis would focus on the relationship between the two sentences, guided by principles like Grice's Cooperative Principle, which assumes speakers are trying to be informative and relevant (Grice, 1975).

> "She left them there after a long day of observing river otters."

The intent here is explanatory. The sentence provides a reason why the glasses are on the bank. This implies a cooperative speaker trying to create a clear narrative for a listener. This understanding of intent is what allows chatbots and voice assistants to provide relevant and helpful responses rather than just literal answers.

3.2.6 Discourse Analysis

Finally, discourse analysis extends beyond individual sentences to examine how they connect to form a coherent, cohesive whole, whether in a conversation, an article, or a story. It analyzes the overall flow of ideas, the thematic structure, and the relationships among the different parts of a text (Schiffrin, 1994). While pragmatics can operate on a single utterance, discourse analysis requires extended communication. In our two-sentence example, discourse analysis would examine the cohesive ties:

> "The researcher's glasses were on the bank. She left them there after a long day of observing river otters."

It would identify that 'she' refers to "the researcher" and 'them' refers to "the glasses." This anaphora resolution is a key function of discourse analysis and is what allows the text to be read as a single, unified narrative rather than two disconnected statements (Hobbs, 1978). This level of analysis is crucial for dialogue systems to maintain context over multiple turns and for tools that summarize long documents by understanding their argumentative or narrative structure.

Discourse Coherence and Rhetorical Structure

Beyond linking entities, discourse analysis is also concerned with **discourse coherence**—what makes a sequence of sentences a unified text rather than a random collection. One of the most influential frameworks for this is **rhetorical structure theory (RST)**. RST proposes that texts are organized hierarchically, with individual clauses or sentences connected by a set of rhetorical relations that describe their purpose relative to one another. These relations are not about syntax, but about the logical and argumentative flow of the text. Some common relations include (Mann & Thompson, 1988):

- **Elaboration:** One part of the text provides more detail about another.

- **Cause/Result:** One part describes the cause or result of another.

- **Contrast:** Two parts of the text present contrasting information.

- **Evidence:** One part provides evidence to support a claim made in another.

Applying this to our running example, we can see that the relationship between the two sentences is not just sequential. The second sentence, "She left them there after a long day of observing river otters," serves a clear rhetorical purpose. It provides a **cause** or **explanation** for the state of affairs described in the first sentence, "The researcher's glasses were on the bank." Recognizing this rhetorical link is a deeper form of understanding. For an application such as text summarization, a system that understands RST can identify the most central claims (the 'nucleus' in RST terms) and distinguish them from supporting details (the 'satellites'), leading to a more intelligent and coherent summary.

3.2.7 Comparative Summary

The following table summarizes the distinct focus of each analysis technique, using our running example to highlight their contributions. This layered approach, in which each level of analysis informs the next, serves as a foundational model for building computational systems that can process and interpret human language (Allen, 1995).

Table 3.1

Comparative Summary

Analysis Type	Focus	Example	Contribution
Lexical Analysis	Individual words and their categories	*bank* (financial institution or riverbank) = Noun; *glass* (eyewear or drinking vessels) = Noun	Identifies words and their potential meanings, but creates ambiguity.
Morphological Analysis	The internal structure of words	*researcher*'s = research + er + 's	Enriches word meaning by analyzing its components.
Syntax Analysis	Sentence structure and grammar	*Subject: The researcher's glasses *Verb: were *Prepositional Phrase: on the bank	Confirms the sentence is grammatically correct, but doesn't resolve meaning.
Semantic Analysis	Literal meaning of words in context	*bank* = riverbank (due to "river otters")	Resolves ambiguity by interpreting the literal meaning based on context.
Pragmatic Analysis	Context and speaker's intended meaning	The 2nd sentence explains why the glasses were on the bank.	Interprets the purpose behind an utterance, going beyond literal meaning.
Discourse Analysis	Coherence and flow across sentences	*She* refers to "the researcher"	Creates a cohesive narrative by linking different parts of the text together.

3.3 The Modern Perspective: Beyond the Pipeline

The six-layered pipeline represents a powerful and essential conceptual model for understanding the different *types* of information that must be extracted from language. It provides a clear framework for the challenges of NLP, from the word level up to the full document level. However, it is crucial to understand that this sequential pipeline does not necessarily reflect how modern, state-of-the-art NLP systems operate.

The rise of end-to-end deep learning models, particularly the transformer architecture that underpins today's large language models, has changed the paradigm. These models do not have separate, distinct modules for lexical, syntactic, and semantic analysis. Instead of passing the output of a parser to a semantic analyzer, a transformer model processes the entire input sequence at once and implicitly learns to capture all these layers of linguistic information simultaneously. The self-attention mechanism allows the model to learn complex relationships between words, regardless of their distance, thereby encoding syntactic dependencies. The deep, multi-layered structure allows the model to build up increasingly abstract representations, effectively capturing semantic meaning. In essence, the lexical, syntactic, and semantic information is all fused together within the rich, contextualized vector embeddings that the model produces for each token. While we can use specialized 'probing' techniques to show that this linguistic information is indeed present in the model's internal representations, the model learns it holistically rather than in a discrete, step-by-step fashion. Therefore, the classic pipeline should be seen as a vital analytical framework for the *problem*, while end-to-end neural networks represent the modern architectural solution.

3.4 Conclusion

This chapter has navigated the essential methods of language analysis, from the granular identification of words to the holistic interpretation of entire texts. We have seen that these techniques are not isolated but form a processing pipeline where each layer builds on the last, working together to resolve ambiguity and construct meaning. The successful execution of this pipeline is what powers today's most sophisticated NLP applications, including virtual assistants, search engines, and machine translation systems.

The future of linguistic research lies in the deeper integration of these analytical layers, particularly to tackle complex phenomena like sarcasm, where semantic meaning and pragmatic intent are in direct opposition. Future work must also focus on adapting these methods to low-resource languages and the ever-evolving slang of digital communication (Pakray et al., 2025; Pasupuleti, 2025). The development of hybrid models that combine the structural strengths of rule-based analysis with the contextual flexibility of deep learning represents a promising frontier (West et al., 2021). As our world becomes more interconnected, the continued advancement of scalable, efficient, and nuanced language analysis will be essential.

3.5 References

Allen, J. F. (1995). *Natural language understanding* (2nd ed.). Benjamin/Cummings Publishing.

Austin, J. L., & Urmson, J. O. (1962). *How to do things with words* (1st ed.). Echo Point Books & Media, LLC.

Chomsky, N. (1957). *Syntactic structures*. Mouton de Gruyter.

Grice, H. P. (1975). *Logic and conversation*. In P. Cole & J. L. Morgan (Eds.), Syntax and semantics, Vol. 3: Speech acts (pp. 41-58). Academic Press.

Gusfield, D. (1997). *Algorithms on strings, trees, and sequences: Computer science and computational biology*. Cambridge University Press. https://doi.org/10.1017/CBO9780511574931

Haspelmath, M., & Sims, A. D. (2010). *Understanding morphology* (2nd ed., 4th impression). Routledge. https://doi.org/10.4324/9780203776506

Hobbs, J. R. (1978). Resolving pronoun references. *Lingua, 44*(4), 311–338. https://doi.org/10.1016/0024-3841(78)90006-2

Jurafsky, D., & Martin, J. H. (2023). *Speech and language processing* (3rd ed.). Prentice Hall.

Lappin, S. (Ed.). (1996). *The handbook of contemporary semantic theory*. Blackwell.

Levinson, S. C. (1983). *Pragmatics*. Cambridge University Press. https://doi.org/10.1017/CBO9780511813313

Mann, W. C., & Thompson, S. A. (1988). Rhetorical structure theory: Toward a functional theory of text organization. *Text-Interdisciplinary Journal for the Study of Discourse, 8*(3), 243–281. https://doi.org/10.1515/text.1.1988.8.3.243

Manning, C. D., & Schütze, H. (1999). *Foundations of statistical natural language processing*. MIT Press.

Navigli, R. (2009). Word sense disambiguation: A survey. *ACM Computing Surveys, 41*(2), 1–69. https://doi.org/10.1145/1459352.1459355

Pakray, P., Gelbukh, A., & Bandyopadhyay, S. (2025). Natural language processing applications for low-resource languages. *Natural Language Processing, 31*(2), 183–197. https://doi.org/10.1017/nlp.2024.33

Pasupuleti, M. K. (2025). Multilingual NLP for low-resource languages using transfer learning. *International Journal of Academic and Industrial Research Innovations(IJAIRI), 5*(5), 452–461. https://doi.org/10.62311/nesx/rphcr7

Schiffrin, D. (1994). Approaches to discourse. Blackwell.

Spencer, A. (1991). *Morphological theory: An introduction to word structure in generative grammar*. Blackwell.

West, P., Bhagavatula, C., Hessel, J., Hwang, J. D., Jiang, L., Bras, R. L., Lu, X., Welleck, S., & Choi, Y. (2021). *Symbolic knowledge distillation: From general language models to commonsense models*. arXiv preprint arXiv:2110.07178.

Winograd, T. (1972). *Understanding natural language*. Academic Press.

Winograd, T. (1980). What does it mean to understand language? *Cognitive Science, 4*(3), 209–241. https://doi.org/10.1207/s15516709cog0403_1

3.6 Glossary

Anaphora Resolution: The task of identifying what a pronoun or other referring expression (e.g., 'she,' 'it,' 'them') refers to in a text.

Discourse Analysis: The study of language beyond the sentence level, focusing on how sentences and utterances connect to form coherent texts or conversations.

Lexical Analysis: The process of identifying and analyzing individual words and the smallest units of meaning in a language.

Morpheme: The smallest meaningful unit in a language, such as a root word, prefix, or suffix.

Morphological Analysis: Examining the structure of words and how they are formed from morphemes.

Natural Language Processing (NLP): A field of computer science and linguistics concerned with the interactions between computers and human languages.

Part-of-Speech (POS) Tag: A label assigned to a token that indicates its grammatical category, such as noun, verb, adjective, or adverb.

Pragmatic Analysis: The examination of how context influences the interpretation of language, considering speaker intent and situational context.

Semantic Analysis: The focus on the meaning of words, phrases, and sentences, interpreting the literal meaning of language in context.

Syntax Analysis (Parsing): The study of the rules and structure of sentences, focusing on how words are arranged to form grammatically correct sentences.

Tokenizing: The process of breaking down text into individual words or lexemes.

Word Sense Disambiguation (WSD): The computational task of determining the correct meaning of a word in a specific context when that word has multiple possible meanings.

3.7 Review Questions and Discussion Topics

1. Given the sentence "The pilots radioed the tower," what lexical ambiguity exists and which level of analysis is required to solve it?

2. Perform a morphological analysis on the word 'deactivated'.

3. Explain why a sentence like "Colorless green ideas sleep furiously" would be considered valid by syntax analysis but meaningless by semantic analysis.

4. If a friend texts you, "I'm fine," after you had an argument, how might a pragmatic analysis differ from a semantic analysis of their message?

5. What cohesive devices does discourse analysis look for to link sentences together? Provide an example.

6. Why is the progression from lexical to discourse analysis often described as a 'pipeline'? What are the potential drawbacks of this layered approach?

7. Ambiguity is the central challenge in NLP. Discuss how lexical, syntactic, and semantic ambiguity differ. Provide your own example of a sentence that contains all three and explain how an NLP system might try to resolve it.

8. Consider the rise of large language models (LLMs). Do these models perform the six types of language analysis discussed here in a distinct, sequential manner, or do they approach the problem differently? Argue for your position.

9. Failures in pragmatic analysis can have significant real-world consequences (e.g., a medical chatbot misinterpreting a patient's description of their symptoms). Discuss the ethical responsibilities of NLP developers in mitigating these risks.

10. Sarcasm is notoriously difficult for computers to understand. Using the layers of analysis from the chapter, explain why it is so challenging. What specific cues (lexical, semantic, pragmatic, etc.) would a system need to identify to detect sarcasm?

11. How might the techniques discussed in this chapter need to be adapted to analyze language from social media platforms, which are characterized by slang, emojis, and a lack of formal sentence structure?

PART II

Core NLP Techniques and Building Blocks

Chapter 4

Natural Language Preprocessing Techniques

4.1 Introduction: The Evolving Role of Preprocessing

The journey to enable machines to comprehend human language is a complex endeavor that stands at the intersection of linguistics, computer science, and artificial intelligence. At the heart of this pursuit lies a critical, often underappreciated, stage: preprocessing. This is the methodical process of cleaning and structuring raw text into a format that a machine can efficiently analyze. For decades, this process was seen as a rigid, sequential pipeline focused on simplifying text for models with limited contextual understanding.

However, the landscape of natural language processing is in a state of rapid, perpetual evolution. The meteoric rise of sophisticated deep learning architectures, particularly the transformer model that underpins modern large language models (LLMs), has instigated a profound paradigm shift. The role of preprocessing has fundamentally transformed. It has evolved from a focus on aggressive text simplification into a strategic, model-aware discipline centered on information preservation and data curation for powerful neural architectures that thrive on linguistic nuance.

This chapter navigates the full spectrum of this evolution. We will begin by critically re-evaluating the foundational normalization pipeline, examining not just the 'what' and 'how' of techniques like stemming and lemmatization, but the more crucial 'why' and 'when.' We will then delve into the subword tokenization revolution, a critical innovation that became a cornerstone of the transformer age, before exploring the new priorities—data quality, deduplication, and bias mitigation—that define preprocessing for modern, web-scale AI. We will conclude with a practical guide summarizing the key differences in preprocessing for classic versus modern models.

See Appendix *A: Technical Deep Dive into Preprocessing.*

4.2 The Normalization Dilemma: Stemming vs. Lemmatization

Text normalization aims to reduce a word's inflectional forms to a common base, consolidating the feature space so that different variations are treated as a single entity. The two most common techniques for this are stemming and lemmatization (Jivani, 2011; Haroon, 2018; Pant et al., 2025).

Figure 4.1

The Normalization Trade-Off: Stemming vs. Lemmatization

Stemming	Lemmatization
Fast & Crude Rule-Based Stripping	Precise Dictionary-Based Look-up
studies → studi	studies → study
running → run	running → run
better → better	better → good
Pros:	**Pros:**
• Very fast • Computationally cheap	• High accuracy • Output is always a valid word
Cons:	**Cons:**
• Often creates non-words • Misses linguistic relationships	• Slower • Requires a dictionary and context (Part-of-Speech)

Stemming is a crude, heuristic-based process that reduces words to their 'stem' by chopping off prefixes or suffixes. The result is not guaranteed to be a valid dictionary word (e.g., 'engines' may become 'engin'). Its primary advantage is speed, which made it a practical choice for early search engines. However, its crudeness often leads to errors like over-stemming (incorrectly merging words like 'university' and 'universe' to the same stem) or under-stemming (failing to merge related words like 'alumnus' and 'alumni') (Porter, 1980).

Lemmatization is a more sophisticated, linguistically informed process. It aims to find the 'lemma', or the canonical dictionary form of a word, by considering its morphological structure and its part-of-speech (POS) in a sentence. For example, it can correctly reduce 'better' to its lemma 'good' or 'mice' to 'mouse'. This accuracy comes at a computational cost, as it is slower and more resource-intensive than stemming (Balakrishnan & Lloyd-Yemoh, 2014).

The motivation for both techniques was to simplify the input for classic models like Bag-of-Words, which treated every word form as a distinct feature. However, this simplification comes at the cost of losing valuable information. Modern transformer-based models are powerful enough to learn the relationship between 'run,' 'ran,' and 'running' automatically. For these models, preserving the original word provides a richer input signal, and stripping away morphological details can be actively detrimental to their performance (Camacho-Collados & Pilehvar, 2017).

4.2.1 A Concrete Comparison of Normalization Techniques

To make the practical differences between these two approaches tangible, consider how they handle a variety of challenging words. The following table compares the output of a popular stemmer (the Porter stemmer) with that of a standard lemmatizer that uses the WordNet lexical database (Miller, 1995).

This side-by-side comparison clearly illustrates the fundamental trade-off. Stemming is a fast, rule-based method that works reasonably well for simple word forms but often fails to handle irregularities and produces non-linguistic tokens. Lemmatization, by leveraging linguistic knowledge and the word's part-of-speech, is far more accurate and always produces a valid dictionary word, but at a higher computational cost. For applications requiring high precision and interpretability, lemmatization is the superior choice; for applications where speed is the absolute priority and some level of error is acceptable, stemming can still be a viable option (Vijayarani et al., 2015).

Table 4.1

Comparison of Normalization Techniques

Original Word	Porter Stemmer Output	Lemmatizer Output (with POS)	Analysis
studies	*studi*	*study*	The stemmer produces a non-word, while the lemmatizer correctly identifies the base noun.
corpora	*corpora*	*corpus*	The stemmer fails to handle the irregular Latin plural, while the lemmatizer correctly normalizes it.
better	*better*	*good*	The stemmer cannot handle this irregular adjective, while the lemmatizer correctly finds the root word 'good.'
meeting	*meet*	*meeting (noun) / meet (verb)*	This highlights the importance of context (POS). As a noun, the lemma is 'meeting'; as a verb, it is 'meet.' The stemmer crudely reduces both to 'meet.'
geese	*gees*	*goose*	Another example of an irregular plural that the stemmer fails to process correctly.
running	*run*	*run*	Both methods succeed on this simple, regular verb form.

4.3 The Subword Revolution: Modern Tokenization

Tokenization, the process of breaking text into smaller units, is the first fundamental step in any NLP pipeline. While classic approaches focused on words, this created an intractable trilemma: massive vocabularies, the out-of-vocabulary (OOV) problem for unknown words, and an inability to recognize morphological relationships. Subword tokenization provided an elegant solution to all three (Mielke et al., 2021).

The core idea is to break words into smaller, meaningful parts. For example, 'unquestionably' might become ['un-,' 'question-,' '-ably']. This approach has profound benefits (Sennrich et al., 2016):

- **Controlled Vocabulary:** A small vocabulary of subwords (e.g., 30,000) can represent a virtually infinite number of words.

- **Elimination of OOV:** Any new word can be constructed from known subwords, preserving its meaning instead of mapping it to a generic <UNK> token.

- **Morphological Awareness:** The model learns the relationship between words like 'running' and 'walking' in two stages. First, the shared subword token ##ing acts as a common signal. Then, by observing this token in similar grammatical contexts across billions of sentences (e.g., often following a verb like 'is' or 'are'), the model learns the semantic function of the morpheme itself, in this case, representing a continuous action.

4.3.1 A Step-by-Step Example of Byte-Pair Encoding (BPE)

To understand how a subword vocabulary is built, let's walk through a simplified example of the byte-pair encoding (BPE) algorithm. Originally a data compression algorithm, BPE was adapted for NLP to address the rare-word problem (Gage, 1994; Sennrich et al., 2016). Assume our entire training corpus is the following set of word counts:

- (low, 5), (lower, 2), (newest, 6), (widest, 3)

Step 1: Initialize Vocabulary and Corpus

First, we break each word into characters and add a special end-of-word symbol (e.g., </w>) to mark word boundaries. The initial vocabulary consists of all unique characters.

- Corpus: l o w </w> (5 times), l o w e r </w> (2 times), n e w e s t </w> (6 times), w i d e s t </w> (3 times)

- Initial Vocabulary: l, o, w, </w>, e, r, n, s, t, i, d

Step 2: Iteratively Merge Most Frequent Pairs

Now, we find the most frequent adjacent pair of tokens in our corpus and merge them into a single new token.

- **Iteration 1:** The pair (e, s) appears most frequently (6 + 3 = 9 times). We merge them to create a new token 'es.'

- ○ **Vocabulary:** ..., es

- ○ **Corpus** now contains n e w es t </w> and w i d es t </w>.

- **Iteration 2:** The pair (es, t) is now the most frequent (9 times). We merge them to create 'est.'

- ○ **Vocabulary:** ..., es, est

- ○ **Corpus** now contains n e w est </w> and w i d est </w>.

- **Iteration 3:** The pair (est, </w>) is the most frequent (9 times). We merge them to create 'est</w>.'

- ○ **Vocabulary:** ..., es, est, est</w>

- ○ **Corpus** now contains n e w est</w> and w i d est</w>.

- **Iteration 4:** The pair (l, o) is now the most frequent (5 + 2 = 7 times). We merge them to create 'lo.'

- ○ **Vocabulary:** ..., es, est, est</w>, lo

- ○ **Corpus** now contains lo w </w> and lo w e r </w>.

This process continues for a predetermined number of merges. The final vocabulary will contain the initial characters plus the learned subword units like es, est, est</w>, and lo. When a new word like 'lowest' is encountered, it can be tokenized into lo, w, est</w>, even if the word itself was never seen during training.

4.3.2 Other Key Subword Algorithms

While BPE is foundational, other influential algorithms offer different strategies for creating subword vocabularies (Schuster & Nakajima, 2012; Kudo & Richardson, 2018; Devlin et al., 2019).

- **WordPiece:** Employed by influential models like BERT, WordPiece is similar to BPE but uses a different merge criterion. Instead of merging the most frequent adjacent pair, it merges the pair that maximizes the likelihood of the training data if the pair were added to the vocabulary. This subtle shift from a frequency-based to a likelihood-based decision often results in a vocabulary better aligned with the language's statistical properties, thereby improving model performance.

- **SentencePiece:** Developed by Google, SentencePiece is a language-agnostic tokenization toolkit that offers a unified approach. Unlike BPE and WordPiece, which often require pre-tokenizing text on whitespace, SentencePiece treats the text as a raw stream of Unicode characters. It also normalizes all whitespace into a single meta-symbol, making it exceptionally robust for languages that do not use spaces to separate words (e.g., Japanese, Chinese). This end-to-end approach makes it ideal for building truly multilingual models.

4.4 The Modern Preprocessing Pipeline for LLMs

The rise of LLMs has shifted the focus of preprocessing from linguistic simplification to data hygiene and curation at a massive scale. The new pipeline prioritizes (Gao et al., 2021):

1. **Data Cleaning and Formatting:** Handling low-level issues like standardizing character encodings (to UTF-8), normalizing whitespace, and removing non-linguistic artifacts like HTML tags or JavaScript code.

2. **Deduplication:** This is one of the most critical steps. Using hashing techniques to remove exact or near-exact duplicate documents prevents the model from overfitting to repeated content (Lee et al., 2022).

3. **Bias and Toxicity Filtering:** Actively curating the training corpus by using classifiers to identify and filter out toxic, hateful, or otherwise undesirable content. This is a crucial step in building responsible and ethical AI systems (Gebru et al., 2021).

4.4.1 Detecting Near-Duplicates at Scale: MinHash and LSH

Removing exact duplicates in a massive dataset is straightforward with standard hashing. The real challenge lies in detecting *near-duplicates*—documents that are almost identical but differ slightly due to formatting, boilerplate text, or minor edits. Comparing every document to every other document results in a complexity of $O(n^2)$, or "order of n-squared," which becomes prohibitively expensive as the dataset grows (see Appendix F, Section F.4). To solve this, large-scale systems use a probabilistic approach based on **min-hashing** and **locality-sensitive hashing**

(LSH) (Broder, 1997).

The process begins by representing each document as a set of 'shingles,' defined as overlapping sequences of five consecutive characters (5-grams). This method slides a fixed-length window across the text to capture local structure, treating whitespace as a valid character alongside letters and numbers. For instance, after normalizing the text by lowercasing and removing punctuation, the phrase "An apple." converts into the following set of shingles: {"an ap", "n app", " appl", "apple"}. The similarity of two documents can then be measured by the Jaccard similarity of their shingle sets. **MinHash** is a clever algorithm that can produce a small 'signature' vector for each document. The magic of MinHash is that the similarity of two documents' signatures is a very good estimate of the Jaccard similarity of their original, much larger shingle sets. This transforms the problem from comparing huge sets to comparing small, fixed-size signatures (Broder et al., 1997).

However, we still need to find which signatures are similar. This is where **locality-sensitive hashing (LSH)** comes in. LSH is a hashing technique with a special property: similar input items (in this case, the MinHash signatures) are highly likely to be sorted into the same category, or 'bucket.' Think of a bucket as a shelf in a library, and the LSH function as a special sorting rule designed to place similar books on the same shelf. By applying this rule to all document signatures, we only need to compare the documents that land in the same bucket for near-duplication. This reduces the problem from an intractable $O(n^2)$ comparison to a much more manageable, near-linear one, making it feasible to find near-duplicates in datasets containing trillions of documents (Indyk & Motwani, 1998).

4.5 No One-Size-Fits-All: The Imperative of Domain-Specific Adaptation

A generic preprocessing pipeline is insufficient for specialized domains. Effective NLP requires strategies tailored to the unique linguistic challenges of the target text (Jurafsky & Martin, 2023).

- **Legal Domain:** Characterized by extreme document length, complex syntax, and specialized jargon. Preprocessing requires robust **optical character recognition (OCR)** for PDFs and specialized **named entity recognition (NER)** models (covered in detail in Chapter 16) trained to identify legal entities such as 'case law,' 'statute,' and 'court' (Chalkidis et al., 2020).

- **Social Media:** Defined by its informality, noise, and multimodality. A successful pipeline must normalize slang and abbreviations, and correctly handle emojis, hashtags, and @mentions, often by translating them into textual representations of their sentiment or meaning (Sarker, 2017).

- **Scientific and Biomedical Domains:** Demands precision and the ability to handle structured information like chemical formulas and gene names. This often involves building custom lexicons (gazetteers) and adapting tokenization strategies to respect domain-specific boundaries (Neumann et al., 2019).

- **Financial Domain:** The language of finance is dense with specific entities and numerical data that require specialized handling. A financial NLP pipeline must be designed to correctly process alphanumeric identifiers like stock tickers (e.g., AAPL, GOOGL) and CUSIPs, which are essential for entity linking. Standard tokenizers might incorrectly split these identifiers. Furthermore, terms like "S&P 500" or "Q4 2023" must be treated as single, meaningful units. The most critical challenge is the accurate parsing and normalization of numerical and currency data (e.g., "\$5.2B," "€1,000,000," "3.5%"), as these values are often the most important pieces of information in a financial document. This requires a preprocessing pipeline with robust regular expressions and rules specifically designed for financial notation (Araci, 2019).

4.6 A Practical Guide to NLP Toolkits

The theoretical principles detailed in the first half of this book, including the mathematics of language models and the core logic of architectures like the transformer, are language-agnostic; the concepts apply universally. However, when moving from theory to implementation—especially in large-scale professional or academic environments—the field has converged on a single programming ecosystem.

Python has become the undisputed standard for modern artificial intelligence and natural language processing. This dominance is due to its simplicity, its strong community, and the high-performance libraries (e.g., Hugging Face, PyTorch,

TensorFlow) built exclusively within its ecosystem.

Therefore, choosing the right tool is critical for implementation. While the Python ecosystem for NLP is vast, the following table compares several of the most prominent and influential libraries, each representing a distinct philosophy (Bird et al., 2009)

Table 4.2

Most Prominent Python Libraries

Feature	NLTK	SpaCy	Hugging Face	Stanza (Stanford)
Primary Use Case	Academic / Research / Education	Production / Application Development	SOTA Deep Learning / Research	Academic Research / Max Accuracy
Performance (Speed)	Slow	Very Fast	Varies (Model Dependent)	Slower than SpaCy
Ease of Use	Moderate	High	High (for pre-trained models)	Moderate
Pre-trained Models	Limited	High-quality pipelines (>60 languages)	Vast hub of SOTA Transformer models	High-accuracy models (>70 languages)
Customization	Very High	Moderate	Very High	High

4.7 The Practitioner's Dilemma: When to Use Which Tool

4.7.1 When to Use Classic, Production-Focused Libraries (like SpaCy)

Classic libraries excel at well-defined, high-volume tasks where speed and efficiency are paramount.

- Use Case 1: High-Speed, Production-Scale Extraction. Using a highly optimized, local library like SpaCy would be orders of magnitude faster and cheaper than making millions of API calls to a large language model.
- Use Case 2: Hybrid Systems. A common and highly effective real-world

architecture is to use a fast library like SpaCy for a first pass over a document to handle 'easy' tasks. Then, only the most complex sentences that require deep reasoning are sent to a more powerful—and expensive—LLM.

4.7.2 When to Use State-of-the-Art LLMs (via Hugging Face)

LLMs are the right choice for tasks that require deep semantic understanding, generation, or complex, multi-step reasoning.

- Use Case 1: Tasks Requiring Nuance and World Knowledge. For complex tasks like abstractive text summarization, answering open-ended questions, or detecting subtle sarcasm, an LLM's vast pre-trained knowledge is indispensable.
- Use Case 2: Zero-Shot or Few-Shot Learning. If you have a unique task but lack a large, labeled dataset to train a specialized model, you can often prompt an LLM to perform the task with just a few examples.

4.8 A Practical Guide: Pipeline Comparison

The central takeaway is that the philosophy of preprocessing has fundamentally changed. The right approach depends entirely on the downstream model. This section serves as the definitive philosophical summary, illustrating the practical outcome of the choice.

4.8.1 Preprocessing for Classic Models (e.g., Naive Bayes, SVM)

- Goal: Simplify & Reduce Text. The objective is a *destructive*, multi-step process that requires reducing the text to its simplest features.
- Key Philosophy: Information is lost (*destructive*) through filtering and stemming because simpler models cannot effectively process the high dimensionality of natural language.
- Process Example (Using spaCy/NLTK): The process yields a simplified list of core content words, such as ['quick,' 'brown,' 'fox,' 'be,' 'jump,' 'lazy,' 'dog'], which is then vectorized for the model. The pipeline steps include tokenization, lowercasing, stop word removal, and stemming or lemmatization.

Figure 4.2

The Evolution of Preprocessing Pipelines

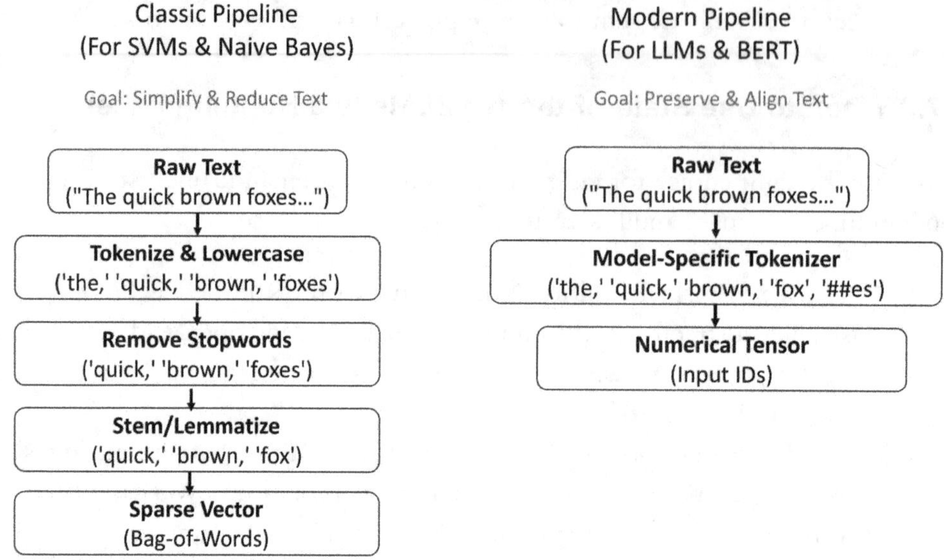

Result: Information is *lost* to help simple models

Result: Information is *preserved* for powerful models

4.8.2 Preprocessing for Pre-trained LLMs (e.g., BERT, GPT)

- Goal: Preserve & Align Text. The objective is to preserve the original linguistic information and perfectly match the format that the pre-trained model expects.

- Key Philosophy: Information is preserved (*non-destructive*) because powerful models (transformers) require and utilize all pieces of context (stop words, morphology, capitalization) to learn deep contextual relationships.

- Process Example (Using Hugging Face): The practitioner loads the model's specific subword tokenizer. This tokenizer performs a single, unified step. The output is a set of numerical tensors, including **input_ids** and an **attention_mask**. No manual stop word removal or lemmatization is performed.

4.9 Assembling a Classic Preprocessing Pipeline

While modern preprocessing is often a single step of tokenizer alignment, understanding the logic of a classic, multi-step pipeline is essential for working with a wide range of NLP models and for building a strong foundational knowledge. The steps are ordered to ensure maximum effectiveness and to prevent later stages from interfering with earlier ones.

A comprehensive classic pipeline generally follows this order:

1. **Text Cleaning**: This is always the first step. It involves removing noise that is irrelevant to linguistic analysis, such as HTML tags, URLs, and special characters. This is done first because these artifacts can interfere with tokenization.

2. **Tokenization**: The cleaned text is broken down into individual words or tokens. This creates the basic units on which all subsequent steps will operate.

3. **Part-of-Speech (POS) Tagging**: After tokenization, each word is tagged with its grammatical category (e.g., noun, verb). This step is crucial because it provides the contextual information needed for accurate lemmatization. POS tagging is covered in Chapter 6.

4. **Lemmatization (or Stemming)**: Words are normalized to their base dictionary form. Lemmatization uses the POS tags from the previous step to guide accuracy, whereas stemming operates on rule-based heuristics without contextual information. This step is done before stop word removal because it is often easier to identify stop words from a canonical list once all words are in their base form.

5. **Stop Word Removal**: Common, low-information words (e.g., 'the,' 'is,' 'a') are filtered out to reduce noise and allow the model to focus on the more meaningful content words.

6. **N-gram Generation:** To recapture some of the local word order context, the preprocessing pipeline treats N-grams (contiguous word sequences like 'New York' or 'not good') as unique, atomic tokens. This preserves limited word order information that would otherwise be lost. This technique is the focus of Chapter 5.

7. **Feature Representation:** Finally, the list of tokens (whether unigrams or N-grams) is converted into a numerical format, such as a bag-of-words or TF-IDF vector. When N-grams are used as the feature set, the resulting Bag-of-Words vector simply counts the frequency of those N-gram sequences, thereby ensuring that the captured word order context is preserved for the downstream machine learning model.

4.10 Conclusion and Future Directions

The discipline of natural language preprocessing has undergone a transformation as profound as that of the NLP models it serves. It has evolved from a set of rigid, language-simplifying heuristics into a dynamic and strategic field of data curation. The classic pipeline, focused on aggressive normalization, has given way to a new philosophy for LLMs: prioritize data quality over simplification. The future trajectory points toward systems that are dynamic, intelligent, and deeply integrated with both the model architecture and the principles of responsible AI, featuring adaptive pipelines and a deepening synergy between data preparation and neural network design.

4.11 References

Araci, D. (2019). *FinBERT: Financial sentiment analysis with pre-trained language models.* arXiv preprint arXiv:1908.10063. https://doi.org/10.48550/arXiv.1908.10063

Balakrishnan, V., & Lloyd-Yemoh, E. (2014). Stemming and lemmatization: A comparison of retrieval performances. *Lecture Notes on Software Engineering, 2*(3), 262–267.

Bird, S., Klein, E., & Loper, E. (2009). *Natural language processing with Python.* O'Reilly Media.

Broder, A. Z. (1997). On the resemblance and containment of documents. *Proceedings of the Compression and Complexity of Sequences* 1997, 21-29. https://doi.org/10.1109/SEQUEN.1997.666900

Broder, A. Z., Glassman, S. C., Manasse, M. S., & Zweig, G. (1997). Syntactic clustering of the Web. *Computer Networks and ISDN Systems, 29*(8-13), 1157–1166. https://doi.org/10.1016/S0169-7552(97)00031-7

Camacho-Collados, J., & Pilehvar, M. T. (2017). *On the role of text preprocessing in neural network architectures: An evaluation study on text categorization and sentiment analysis.* arXiv preprint arXiv:1707.01780. https://doi.org/10.48550/arXiv.1707.01780

Chalkidis, I., Fergadiotis, M., Malakasiotis, P., Aletras, N., & Androutsopoulos, I. (2020). LEGAL-BERT: The muppets straight out of law school. *Findings of the Association for Computational Linguistics: EMNLP 2020,* 2898–2904. https://doi.org/10.18653/v1/2020.findings-emnlp.261

Devlin, J., Chang, M. W., Lee, K., & Toutanova, K. (2019). BERT: Pre-training of deep bidirectional transformers for language understanding. *Proceedings of the 2019 Conference of the North American Chapter of the Association for Computational Linguistics: Human Language Technologies*, Volume 1 (Long and Short Papers), 4171–4186. https://doi.org/10.18653/v1/N19-1423

Gage, P. (1994). A new algorithm for data compression. *The C Users Journal*, 12(2), 23-38.

Gao, L., Biderman, S., Black, S., Golding, L., Hoppe, T., Foster, C., ... & LeGresley, P. (2021). *The Pile: An 800GB dataset of diverse text for language modeling.* arXiv preprint arXiv:2101.00027. https://doi.org/10.48550/arXiv.2101.00027

Gebru, T., Morgenstern, J., Vecchione, B., Vaughan, J. W., Wallach, H., Daumé III, H., & Crawford, K. (2021). Datasheets for datasets. *Communications of the ACM*, *64*(12), 86–92. https://doi.org/10.1145/3458723

Haroon, M. (2018). Comparative analysis of stemming algorithms for web text mining. *International Journal of Modern Education and Computer Science, 10*(9), 20–25. https://doi.org/10.5815/ijmecs.2018.09.03

Indyk, P., & Motwani, R. (1998). Approximate nearest neighbors: Towards removing the curse of dimensionality. *Proceedings of the Thirtieth Annual ACM Symposium on Theory of Computing*, 604–613. https://doi.org/10.1145/276698.276876

Jivani, A. G. (2011). A comparative study of stemming algorithms. *International Journal of Computer Technology and Applications, 2*(6), 1930–1938.

Jurafsky, D., & Martin, J. H. (2023). *Speech and language processing* (3rd ed.). Prentice Hall.

Kudo, T., & Richardson, J. (2018). SentencePiece: A simple and language independent subword tokenizer and detokenizer for neural text processing. *In Proceedings of the 2018 Conference on Empirical Methods in Natural Language Processing: System Demonstration*s (pp. 66–71). Association for Computational Linguistics. https://doi.org/10.18653/v1/D18-2012

Lee, K., Ippolito, D., Nystrom, A., Zhang, C., Eck, D., Callison-Burch, C., & Carlini, N. (2022). Deduplicating training data makes language models better. *Proceedings of the 60th Annual Meeting of the Association for Computational Linguistics* (Volume 1: Long Papers), 8424–8445. https://doi.org/10.18653/v1/2022.acl-long.577

Mielke, S. J., Cotterell, R., Eisner, J., & Blasi, D. E. (2021). *Between words and characters: A brief history of subword segmentation in natural language processing.* arXiv preprint arXiv:2112.10508. https://doi.org/10.48550/arXiv.2112.10508

Miller, G. A. (1995). WordNet: A lexical database for English. *Communications of the ACM, 38*(11), 39–41. https://doi.org/10.1145/219717.219748

Neumann, M., King, D., Beltagy, I., & Ammar, W. (2019). *ScispaCy: Fast and robust models for biomedical natural language processing.* arXiv preprint arXiv:1902.07669. https://doi.org/10.48550/arXiv.1902.07669

Pant, V. K., Sharma, R., & Kundu, S. (2025). *An overview of stemming and lemmatization techniques.* In P. Rattan, A. Kumar, A. Sharma, P. Chopra, & G. Sharma (Eds.), Advances in Networks, Intelligence and Computing (1st ed., pp. 308–321). CRC Press. https://doi.org/10.1201/9781003430421-31

Porter, M. F. (1980). An algorithm for suffix stripping. *Program, 14*(3), 130–137. https://doi.org/10.1108/eb046814

Sarker, A. (2017). A customizable pipeline for social media text normalization. *Social Network Analysis and Mining, 7*(1), 45. https://doi.org/10.1007/s13278-017-0464-z

Schuster, M., & Nakajima, K. (2012). Japanese and Korean voice search. In *2012 IEEE International Conference on Acoustics, Speech and Signal Processing (ICASSP)* (pp. 5149–5152). IEEE. https://doi.org/10.1109/ICASSP.2012.6289079

Sennrich, R., Haddow, B., & Birch, A. (2016). Neural machine translation of rare words with subword units. *Proceedings of the 54th Annual Meeting of the Association for Computational Linguistics* (Volume 1: Long Papers), 1715–1725. https://doi.org/10.18653/v1/P16-1162

Vijayarani, S., Ilamathi, M. J., & Nithya, M. (2015). Preprocessing techniques for text mining - an overview. *International Journal of Computer Science & Communication Networks, 5*(1), 7–16.

4.12 Glossary

Byte-Pair Encoding (BPE): A subword tokenization algorithm that iteratively merges the most frequent pair of adjacent tokens in a corpus.

Deduplication: The process of identifying and removing duplicate or near-duplicate entries from a dataset to improve data quality and prevent model overfitting.

Jaccard Similarity: A statistic used for gauging the similarity of two sets, calculated as the size of the intersection divided by the size of the union of the sets.

Lemmatization: A linguistically informed process of reducing a word to its base or dictionary form (lemma).

Locality-Sensitive Hashing (LSH): A hashing technique designed to group similar items into the same 'buckets' with high probability, used for efficient near-duplicate detection.

MinHash: An algorithm used to quickly estimate the Jaccard similarity between two sets, often by creating a compressed 'signature' or fingerprint of each set.

Named Entity Recognition (NER): A subtask of information extraction that seeks to locate and classify named entities in text into pre-defined categories such as person names, organizations, and locations.

Optical Character Recognition (OCR): The technology used to convert images of typed, handwritten, or printed text into machine-encoded text.

Out-of-Vocabulary (OOV): The problem where a model encounters words during inference that were not present in its training vocabulary.

SentencePiece: A language-agnostic tokenization toolkit that treats text as a raw stream of Unicode characters, making it ideal for multilingual models.

Shingles: Short, overlapping sequences of characters or words (n-grams) used to represent a document as a set for similarity comparisons.

Stemming: A heuristic process of removing suffixes and prefixes from words to reduce them to a common root form.

Subword Tokenization: A class of tokenization algorithms that break words into smaller, meaningful parts, resolving the issues of large vocabularies and OOV words.

WordPiece: A subword tokenization algorithm used by BERT that merges token pairs based on which merge maximizes the likelihood of the training data.

4.13 Review Questions and Discussion Topics

1. **Modern Pipeline Design:** You are tasked with preprocessing a massive dataset scraped from the web to train a new LLM. Based on this chapter, what would be your top three priorities for the preprocessing pipeline? Justify your choices.

2. **Tokenization Trade-offs:** Compare and contrast byte-pair encoding (BPE) and WordPiece. Why might a model like BERT, which is focused on deep contextual understanding, prefer WordPiece's likelihood-based approach over BPE's frequency-based approach?

3. **Domain Adaptation:** You have a sentiment analysis model that performs well on product reviews but fails on legal documents. What specific characteristics of legal text, as discussed in the chapter, are likely causing the failure? What preprocessing steps would you take to adapt your system?

4. **The Role of Simplification:** While the chapter argues that aggressive normalization is often harmful for LLMs, can you think of a specific NLP task or a resource-constrained scenario where using stemming or heavy stop word removal might still be a valid and effective strategy?

Chapter 5

N-grams: From Foundations to the Transformer Era

5.1 Introduction: The Power of Word Sequences

Imagine trying to predict the next word someone will type on their phone. Or think about how a search engine understands the nuances of your query. At the heart of many language understanding tasks lies a fundamental concept: the sequence of words. N-grams, contiguous sequences of 'n' items (typically words), provide a powerful and intuitive way to capture local context within text. They allow us to move beyond individual words and consider the relationships between adjacent terms, forming the bedrock for various natural language processing (NLP) applications (Jurafsky & Martin, 2023). This chapter will delve into the world of N-grams, exploring their definition, diverse applications, inherent limitations, and crucially, their role as a stepping stone towards the sophisticated transformer architectures that power much of modern AI.

5.2 What are N-grams? The Building Blocks of Language Understanding

At its core, an N-gram is a sequence of 'n' consecutive items from a given text or speech. These items can be characters, syllables, or, most commonly in NLP, words. The 'n' in N-gram indicates the number of items in the sequence, a concept rooted in Claude Shannon's foundational work on information theory, which modeled language as a sequence of symbols (Shannon, 1948). For instance, consider the sentence: "The quick brown fox jumps."

- **Unigrams (n=1):** 'The,' 'quick,' 'brown,' 'fox,' 'jumps'

- **Bigrams (n=2):** "The quick," "quick brown," "brown fox," "fox jumps"

- **Trigrams (n=3):** "The quick brown," "quick brown fox," "brown fox jumps"

And so on…

Figure 5.1

Deconstructing N-grams

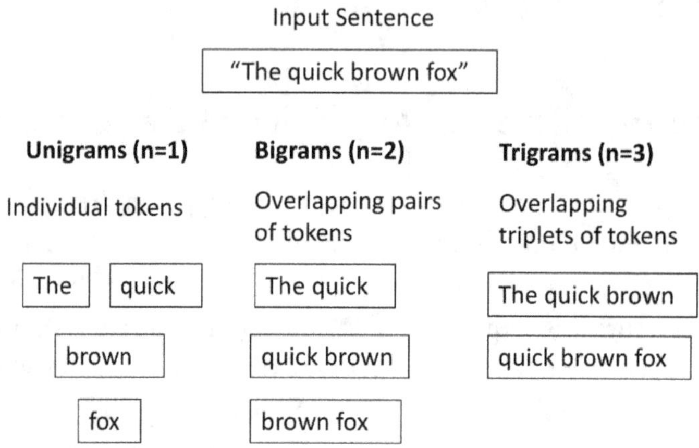

The choice of 'n' depends heavily on the specific task. Smaller 'n' values capture very local context, while larger 'n' values consider broader sequential information but introduce significant computational challenges. This trade-off between context and computability has been a central theme in the history of statistical language modeling (Manning & Schütze, 1999).

5.3 The Utility of N-grams: Applications Across NLP

Despite their simplicity, N-grams have proven incredibly useful across a wide spectrum of NLP tasks, largely because they provide a direct way to estimate the probability of a word given its local context (Jurafsky & Martin, 2023).

5.3.1 Language Modeling and Generation

- **Language Modeling (The Core Capability):** This is the quintessential application and the foundational capability for many other tasks. A language model's primary goal is to assign a probability to a sequence of text, effectively measuring how natural or likely a sentence is. Its most direct application is predicting the *single most likely next word* given a preceding sequence. This is the foundation of predictive text keyboards and speech recognition systems (Jelinek, 1990). For example, if we see "peanut butter and," a language model predicts 'jelly' as the probable next word.

- **Text Generation (A Downstream Application):** This is a creative task that *uses* a language model iteratively. While language modeling focuses on the probability of the very next word, text generation uses that capability repeatedly to build longer, novel sequences of text from a starting prompt. The process is iterative: the model predicts a word, appends it to the sequence, and then uses this new, longer sequence as the context to predict the next word. While N-gram-based generation often lacks long-range coherence, it can be useful for creating short, contextually relevant phrases (Manning & Schütze, 1999).

5.3.2 Classification and Information Retrieval

N-grams proved invaluable in the statistical era for categorizing and retrieving documents based on local word sequences.

- **Text Classification:** The frequency of certain N-grams can be indicative of the topic, sentiment, or author of a text. For sentiment analysis, N-grams often function as powerful, isolated features, proving their value in distinguishing emotional tone (Pang et al., 2002; Georgieva-Trifonova & Daraku, 2021). For instance, a high frequency of negative sentiment words within bigrams might suggest a negative review.

- **Machine Translation:** While largely superseded by neural models, early statistical machine translation (SMT) systems utilized N-gram models to ensure the fluency of the target language output, scoring potential translations based on how 'probable' their word sequences were in the target language (Koehn et al., 2003; Sennrich, 2015).

5.3.3 Text Utility and Error Correction

- **Spell Checking and Error Correction:** N-grams can help identify and correct spelling errors. A less frequent or unseen N-gram might indicate a potential error. By comparing it to similar, more frequent N-grams (with single word substitutions, insertions, or deletions), a spell checker can suggest corrections based on the noisy channel model (Brill & Moore, 2000; Boinski et al., 2018).

5.4 The Mathematics of N-gram Language Models

The core task of a language model is to assign a probability to a sequence of words. The probability of a sequence $W=(W_1, W_2, ..., W_k)$ can be decomposed using the **chain rule of probability**:

Equation 5.1

Chain Rule of Probability

$$P(W) = P(w_1) \times P(w_2|w_1) \times P(w_3|w_1, w_2) \times \cdots \times P(w_k|w_1, \ldots, w_{k-1})$$

This formula states that the probability of the entire sequence is the product of the conditional probabilities of each word given all the words that came before it. However, calculating these long conditional probabilities is computationally intractable and suffers from extreme data sparsity. It is highly unlikely we would ever see the exact sequence "The quick brown fox" in our training data to predict the word 'jumps' (Brown et al., 1992). See Appendix B for more information about the chain rule of probability and its application.

To make this problem feasible, N-gram models apply the **Markov assumption**, which posits that the probability of a word depends only on a limited history of preceding words. For an N-gram model, we assume the probability of a word depends only on the previous n-1 words. For a bigram model (n=2), this simplifies the problem dramatically: the probability of a word depends only on the single word that precedes it. See Appendix C for more information about the Markov assumption for a bigram model.

Equation 5.2

Markov Assumption $\quad P(w_i|w_1, \ldots, w_{i-1}) \approx P(w_i|w_{i-1})$

Note: This reads: "is approximately equal to the probability of the word W_i, given only the single word that came immediately before it. "**Example:** Using the approximation, we say P(sat | The, cat) ≈ P(sat | cat). We completely ignore the word 'The.'

Figure 5.2

Theoretical Problem: Full Chain Rule

Theoretically correct, but computationally impossible.

The quick brown fox...

P(fox | The, quick, brown) = ?

Limitation: Requires calculating the probability of a word given an exponentially long history. The model would need to have seen the exact phrase "The quick brown" many times to make a prediction, which is highly improbable.

Figure 5.3

The Markov Assumption: A "Great Simplification"

We simplify by assuming a limited "memory."

Bigram Model (n=2)

P(fox | The, quick, brown) ≈ P(fox | brown)

(Memory = 1 word)

Trigram Model (n=3)

P(fox | The, quick, brown) ≈ P(fox | quick, brown)

(Memory = 2 words)

Advantage: This is now a simple, solvable problem. We only need to count word pairs (bigrams) or triplets (trigrams), which is computationally feasible.

With this assumption, we can estimate these probabilities from a corpus using **maximum likelihood estimation** (MLE). The MLE for a bigram probability is simply the count of the bigram divided by the count of its prefix unigram:

Equation 5.3

Maximum Likelihood Estimation

$$P(w_i|w_{i-1}) = \frac{\text{count}(w_{i-1}, w_i)}{\text{count}(w_{i-1})}$$

For example, to calculate the probability of 'fox' following 'brown' in a corpus, we would count every occurrence of the bigram "brown fox" and divide it by the total number of times the word 'brown' appears. If "brown fox" appears 800 times and 'brown' appears 1000 times, then P(fox | brown)=800/1000=0.8. This simple, count-based approach forms the statistical foundation of N-gram language modeling (Manning & Schütze, 1999). See Appendix D for more information about MLE and the application.

5.5 The Limitations of N-grams: Where the Simple Approach Falls Short

Despite their successes, N-gram models suffer from several fundamental limitations that ultimately necessitated the shift toward neural approaches (Bengio et al., 2003; Kumar & Thirumaran, 2024).

- **Data Sparsity:** As the value of 'n' increases, the number of possible N-grams grows exponentially. Consequently, even very large training corpora may not contain all possible N-grams, leading to unseen sequences (the "zero-frequency problem"). This sparsity makes it difficult to accurately estimate probabilities for longer sequences. This issue led to the development of numerous 'smoothing' techniques designed to assign some small probability to unseen events, a critical area of research in statistical NLP (Chen & Goodman, 1998).

- **Inability to Capture Long-Range Dependencies:** N-gram models only consider a limited context of 'n-1' preceding words. They struggle to capture relationships between words that are far apart in a sentence or across multiple sentences. For example, in the sentence "The dog, which my neighbor adopted last week and is known for its playful nature, loves to bark loudly," a trigram model trying to predict the word after 'bark' would likely only consider "loves to." It would have no direct way of knowing that 'dog' (much earlier in the sentence) is the subject performing the action. This inability to model long-term dependencies is a well-documented weakness of this class of models (Bengio et al., 2003).

- **Ignoring Semantic Similarity:** N-gram models treat words as discrete symbols. They do not inherently understand the semantic similarity between words. For instance, 'car' and 'automobile' would be considered completely different tokens, even though they have very similar meanings. This can lead to poor generalization if one word appears frequently in the training data but its synonyms do not (Manning & Schütze, 1999).

5.5.1 The Zero-Frequency Problem and Smoothing

The data sparsity issue leads to a critical mathematical failure known as the zero-frequency problem. If a specific N-gram (e.g., "San Francisco Giants") never appeared in our training corpus, its count would be zero. According to the MLE formula, the probability of this N-gram would also be zero. This is catastrophic because, according to the chain rule, the probability of any longer sequence of text containing this unseen N-gram will also become zero. The model would thus assign a probability of zero to a perfectly valid sentence it simply hasn't encountered before. To solve this, a class of statistical techniques called **smoothing** (or discounting) was developed. The core idea of smoothing is to take a small amount of probability mass from the N-grams we *have* seen and redistribute it to the N-grams we *haven't* seen (Gale & Sampson, 1995).

The simplest and most foundational of these techniques is **laplace smoothing**, also known as **add-one smoothing**. It works by adding one to every N-gram count before calculating the probabilities. The formula for a smoothed bigram probability becomes:

Equation 5.4

Laplace Smoothing

$$P_{\text{Laplace}}(w_i | w_{i-1}) = \frac{\text{count}(w_{i-1}, w_i) + 1}{\text{count}(w_{i-1}) + V}$$

Here, V represents the size of the entire vocabulary. By adding one to the numerator, we ensure that no N-gram has a zero probability. The vocabulary size V is added to the denominator to ensure that the probabilities still sum to one. While simple and effective at eliminating zero probabilities, Add-One smoothing is a blunt instrument. It often gives far too much probability mass to the vast number of unseen events, and more sophisticated techniques like Good-Turing, Kneser-Ney, and Katz's backoff were developed to provide more nuanced and accurate probability estimates (Chen & Goodman, 1998). See Appendix E for more details about Laplace Smoothing and its application.

5.6 Evaluating Language Models: Perplexity

To compare the performance of different language models, we need a reliable evaluation metric. The standard intrinsic metric for language models is **perplexity**. Perplexity is a measure of how well a probability model predicts a sample. In the context of NLP, it measures how 'surprised' or 'perplexed' a language model is by a previously unseen test set. A lower perplexity score indicates that the model is better at predicting the sample, and is therefore a better language model (Bahl et al., 1983).

Mathematically, the perplexity of a language model on a test set $W = (W_1, W_2, ..., W_N)$ is the inverse probability of the test set, normalized by the number of words:

Equation 5.5

Perplexity

$$\text{Perplexity}(W) = P(w_1, w_2, \ldots, w_N)^{-1/N}$$

Perplexity indicates the average number of equally likely next words the model can choose from at each step in a sequence. For instance, a perplexity score of 150 means the model performs as if it were randomly selecting from 150 equally likely words at every position. This measure is much more interpretable than the raw probability on the test set, which is often very small. A lower perplexity score (e.g., 120) is always preferable to a higher one (e.g., 150), as it indicates the model is more confident and makes predictions from a smaller, more certain set of choices. Perplexity is closely related to cross-entropy, another information-theoretic measure. For a language model, minimizing perplexity is equivalent to minimizing cross-entropy. When developing a new language model, researchers will typically train different versions of the model on a training set and then compare their perplexity scores on a held-out test set to determine which model is superior (Jurafsky & Martin, 2023).

5.7 Contrasting Statistical and Neural Language Models

Despite their limitations, the concept of capturing local word order inherent in N-grams remains valuable. Modern deep learning models, particularly those used in NLP, often incorporate mechanisms that implicitly or explicitly capture similar local contextual information (Goldberg, 2017). However, the shift from count-based statistical models to neural language models represented a major leap forward, as the neural approach inherently solves two of the N-gram model's most significant weaknesses (Bengio et al., 2003): semantic similarity and the sparsity problem.

A neural language model works by taking a sequence of preceding words as input, converting them into dense vector representations (word embeddings), processing these vectors through one or more neural network layers, and finally producing a probability distribution over the entire vocabulary for the next word. This architecture provides two key advantages over a statistical N-gram model. First, it solves the problem of **semantic similarity**. Because the learned embeddings for semantically similar words like 'car and 'automobile' are close to each other in the vector space, the model can naturally generalize. If the model has learned from the phrase "the driver of the car," it can infer that "the driver of the automobile" is also a probable phrase, even if it has never seen that exact trigram in the training data. An N-gram model cannot do this (Mikolov et al., 2013).

Second, neural models are far more robust to the **sparsity problem**. Instead of relying on discrete counts of specific N-grams, the model learns a continuous function over the distributed representations of words. This allows it to estimate the probability of a novel combination of words much more effectively than a statistical model that would have to fall back on complex smoothing techniques. In essence, neural language models learn a much smoother and more generalizable representation of language, which is why they have become the standard for modern NLP tasks (Bengio et al., 2003).

5.8 Conclusion: From Simple Sequences to Powerful Architectures

N-grams provide a foundational understanding of how local word order can be captured and utilized in NLP. While they face limitations in dealing with data sparsity, long-range dependencies, and semantic understanding, their simplicity and effectiveness in various tasks made them a cornerstone of early NLP research and applications. The move towards integrating N-gram insights into neural network architectures, and ultimately the development of transformer networks with their powerful attention mechanisms, represents a significant leap forward in our ability to model and understand human language. The journey from the basic concept of counting word sequences to the complex machinery of modern language models highlights the continuous evolution and innovation within the field of computer science and artificial intelligence.

5.9 References

Bahl, L. R., Jelinek, F., & Mercer, R. L. (1983). A maximum likelihood approach to continuous speech recognition. *IEEE Transactions on Pattern Analysis and Machine Intelligence, 5*(2), 179–190. https://doi.org/10.1109/TPAMI.1983.4767370

Bengio, Y., Ducharme, R., Vincent, P., & Jauvin, C. (2003). A neural probabilistic language model. *Journal of Machine Learning Research, 3*, 1137–1155.

Boinski, T., Zimnicki, A., Kujawski, J., & Draszawka, K. (2018). Evaluating asymmetric N-grams as spell-checking mechanism. *Proceedings of the 2018 11th International Conference on Human System Interaction (HSI)*, 356–362.

Brill, E., & Moore, R. C. (2000). An improved error model for noisy channel spelling correction. *Proceedings of the 38th Annual Meeting of the Association for Computational Linguistics*, 286–293. https://doi.org/10.3115/1075218.1075255

Brown, P. F., Della Pietra, S. A., Della Pietra, V. J., Mercer, R. L., & Nadas, L. (1992). An estimate of an upper bound for the entropy of English. *Computational Linguistics, 18*(1), 31–40.

Chen, S. F., & Goodman, J. (1998). *An empirical study of smoothing techniques for language modeling* (Technical Report TR-10-98). Harvard University.

Gale, W. A., & Sampson, G. (1995). Good-Turing frequency estimation without tears. *Journal of Quantitative Linguistics, 2*(3), 217–237. https://doi.org/10.1080/09296179508590051

Georgieva-Trifonova, T., & Duraku, M. (2021). Research on N-grams feature selection methods for text classification. *IOP Conference Series. Materials Science and Engineering, 1031*(1), 12048. https://doi.org/10.1088/1757-899X/1031/1/012048

Goldberg, Y. (2017). *Neural network methods for natural language processing.* Springer Nature. https://doi.org/10.1007/978-3-031-02165-7

Jelinek, F. (1990). Self-organized language modeling for speech recognition. In A. Waibel & K. F. Lee (Eds.), *Readings in speech recognition* (pp. 450–506). Morgan Kaufmann.

Jurafsky, D., & Martin, J. H. (2023). *Speech and language* processing (3rd ed.). Pearson.

Koehn, P., Och, F. J., & Marcu, D. (2003). Statistical phrase-based translation. *Proceedings of the 2003 Conference of the North American Chapter of the Association for Computational Linguistics on Human Language Technology, 1*, 48–54.

Kumar, R. K., & Thirumaran, S. (2024). Enhancing automatic English word analysis and prediction using higher-order N-gram models. *2024 International Conference on Science Technology Engineering and Management (ICSTEM)*, 1–7. https://doi.org/10.1109/ICSTEM61137.2024.10560953

Manning, C. D., & Schütze, H. (1999). *Foundations of statistical natural language processing.* MIT Press.

Mikolov, T., Chen, K., Corrado, G., & Dean, J. (2013). *Efficient estimation of word representations in vector space.* arXiv preprint. https://arxiv.org/abs/1301.3781

Pang, B., Lee, L., & Vaithyanathan, S. (2002). Thumbs up? Sentiment classification using machine learning techniques. *Proceedings of the 2002 Conference on Empirical Methods in Natural Language Processing, 10*, 79–86. https://doi.org/10.3115/1118693.1118704

Sennrich, R. (2015). Modelling and optimizing on syntactic N-grams for statistical machine translation. *Transactions of the Association for Computational Linguistics, 3*, 169–182. https://doi.org/10.1162/tacl_a_00131

Shannon, C. E. (1948). A mathematical theory of communication. *The Bell System Technical Journal, 27*(3), 379–423. https://doi.org/10.1002/j.1538-7305.1948.tb01338.x

5.10 Glossary

Bigram: A sequence of two contiguous items (typically words) from a sample of text or speech.

Chain Rule of Probability: A fundamental rule of probability theory used to calculate the probability of a sequence of events.

Data Sparsity: A problem in statistical models where many possible events (like specific N-grams) do not appear in the training data, leading to incorrect zero-probability estimates.

Laplace Smoothing (Add-One Smoothing): A simple technique to address the zero-frequency problem by adding one to all N-gram counts.

Markov Assumption: A simplifying assumption made in sequence models which posits that the probability of the current event depends only on a limited number of previous events.

Maximum Likelihood Estimation (MLE): A common-sense method for estimating the probabilities in a statistical model. The core idea is to find the values that make our observed data the most likely. For N-gram models, this means the probability of a word sequence is simply its frequency in the training text.

N-gram: A contiguous sequence of 'n' items (typically words) from a given sample of text or speech.

Natural Language Processing (NLP): A field of computer science focused on enabling computers to understand, interpret, and manipulate human language.

Neural Language Models: Language models based on neural networks that learn distributed representations (embeddings) of words, allowing them to handle semantic similarity and data sparsity more effectively than statistical models.

Perplexity: The standard intrinsic evaluation metric for a language model, measuring how well it predicts a test set. A lower perplexity indicates a better model.

Sentiment Analysis: The process of determining the emotional tone or sentiment expressed in a piece of text.

Smoothing: A class of techniques used in statistical language models to assign a non-zero probability to unseen events, addressing the zero-frequency problem.

Text Classification: The task of categorizing text into predefined classes based on its content.

Text Generation: The process of creating new text based on input data or a given model.

Token: A single unit of text or speech, such as a word, punctuation mark, or number.

Trigram: A sequence of three contiguous items (typically words) from a sample of text or speech.

Unigram: A single item (typically a word) from a sample of text or speech.

Zero-Frequency Problem: The issue in statistical models where an event that did not occur in the training data is assigned a probability of zero, which can cause mathematical problems when calculating the probability of larger sequences.

5.11 Review Questions and Discussion Topics

1 Provide the unigrams, bigrams, and trigrams for the sentence: "Artificial intelligence is rapidly changing the world."

2 Explain the problem of data sparsity in the context of 5-grams. Give a concrete example of a 5-gram that might be very sparse in a general English corpus.

3 Describe a scenario where an N-gram language model might struggle to predict the correct next word due to its inability to capture long-range dependencies.

4 How does the concept of 'attention' in transformer networks address the limitations of N-gram models in handling long-range dependencies?

5 Consider the bigrams from the sentence: "The cat sat on the mat." How might a bigram language model use these to predict the next word after 'sat'? What are the potential limitations in this prediction?

6 Discuss the trade-offs between using smaller 'n' values versus larger 'n' values in N-gram models. Consider factors such as data sparsity, the ability to capture context, and computational cost. How might the optimal choice of 'n' vary depending on the specific NLP task?

Chapter 6

Part-of-Speech Tagging

6.1 Introduction to Part-of-Speech (POS) Tagging

Part-of-speech (POS) tagging is a fundamental process in natural language processing (NLP) that involves assigning a specific grammatical category—such as noun, verb, adjective, or adverb—to each word in a text. This linguistic annotation is crucial, as it provides the syntactic scaffolding required for sophisticated computational analysis of language, transforming raw text into structured data that machines can interpret.

The core challenge in POS tagging stems from the inherent ambiguity of natural language. A single word can serve multiple grammatical functions depending on its context. For example, the word 'run' can be a noun ("She went for a long run") or a verb ("They run every morning"). Similarly, in the sentence "The sailor dogs the hatch," the word 'dogs' functions as a verb meaning "to fasten securely," not as a plural noun. POS tagging methodologies are designed to resolve precisely this type of ambiguity by analyzing a word's definition in conjunction with its contextual relationships.

The granularity of POS tagging can vary significantly. Basic systems might use a small set of tags, while more comprehensive systems, particularly for English, may distinguish between 50 to 150 separate parts of speech, incorporating additional linguistic features such as tense, case, and grammatical gender. By providing this detailed syntactic information, POS tagging serves as an essential preprocessing step for a wide array of advanced NLP tasks, including named entity recognition (NER), machine translation, sentiment analysis, and information retrieval.

Figure 6.1

The Core Task of Part-of-Speech Tagging

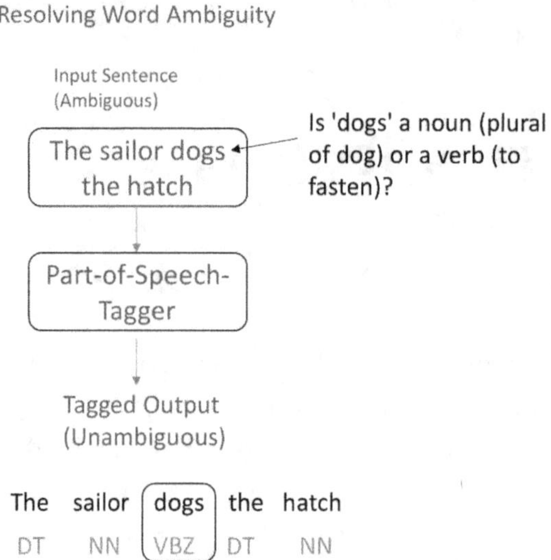

6.2 The Distinction: POS Tagging vs. POS Tagger

Within computational linguistics, it is important to distinguish between the process and the tool. This distinction clarifies the relationship between a linguistic task and its computational implementation, a common paradigm in the field.

- **Part-of-Speech Tagging** refers to the process or task of assigning grammatical categories to words. It is the conceptual act of analyzing a word's meaning and context to determine its function within a sentence (Jurafsky & Martin, 2023).

- **A Part-of-Speech Tagger** is the tool or algorithm used to perform the tagging process. It is the computational implementation—whether a set of linguistic rules or a machine learning model—designed to automate the identification and assignment of POS tags to words in a text (Manning, 2011).

In essence, POS tagging is the linguistic task, while a POS tagger is the computational mechanism that executes it.

6.3 Core Applications of POS Tagging

By categorizing words into their grammatical classes, POS tagging provides crucial syntactic information that enhances the performance of numerous NLP applications. This process of disambiguation and structural annotation is a key enabling step for more complex, meaning-oriented tasks. Key applications can be grouped as follows:

6.3.1 Syntactic and Semantic Analysis

- **Text Parsing:** POS tags are essential for both dependency parsing (identifying grammatical relationships like subject-verb-object) and constituency parsing (breaking sentences into phrases like noun phrases and verb phrases) (Jurafsky & Martin, 2023).

- **Word Sense Disambiguation (WSD):** Tags provide vital contextual clues to determine the correct meaning of a word with multiple senses. For example, identifying 'bank' as a noun versus a verb helps disambiguate its meaning (Navigli, 2009; Dubey, 2024).

- **Coreference Resolution:** Tagging helps identify which words (especially pronouns) refer to the same entity. Identifying 'he' as a pronoun is the first step in linking it back to a specific entity like 'John' (Hobbs, 1978; Liu et al., 2023).

6.3.2 Information Systems

- **Information Extraction and Retrieval:** POS tagging helps pinpoint key information, such as names of people, locations, and organizations, from unstructured text. Search engines use tags to better understand user queries, leading to more relevant results (Grishman, 2019).

- **Question Answering (QA) Systems:** Tags help QA systems analyze the structure of a question to determine what kind of information is being sought (e.g., a person, location, or date) (Moldovan et al., 2003; Ojokoh & Adebisi, 2019).

- **Text Summarization:** By identifying the most important words (e.g., key nouns and verbs), POS tagging helps generate concise, relevant summaries of documents (Nenkova & McKeown, 2012; Sharma & Sharma, 2023).

6.3.3 Advanced NLP Tasks

- **Named Entity Recognition (NER):** POS tagging is fundamental to NER systems, as it helps distinguish proper nouns (e.g., 'Google') from common nouns, a critical step in identifying specific entities (Nadeau & Sekine, 2007; Reddy et al., 2023).

- **Machine Translation (MT):** Accurate translation relies on analyzing the grammatical roles of words in the source language to ensure correct alignment and structure in the target language (Koehn, 2009; Das et al., 2025).

- **Sentiment Analysis:** Tagging contributes to more nuanced sentiment analysis by helping identify words that are strong indicators of sentiment, particularly adjectives and adverbs (Pang & Lee, 2008; Keith et al., 2019).

6.3.4 Language and Speech Technologies

- **Speech Synthesis and Text-to-Speech (TTS):** In TTS systems, tags help determine the correct pronunciation and intonation. The word 'read' is pronounced differently depending on whether it is tagged as a present tense verb (/riːd/) or a past tense verb (/rɛd/) (Taylor, 2009; Kaur & Singh, 2023).

- **Grammar Checking:** Grammar checkers use POS tagging to identify incorrect word usage or grammatical errors by analyzing the expected syntactic structure of a sentence (Heidorn, 2000; Gholami & Feizi, 2023; Chiche & Yitagesu, 2022).

6.4 Methodologies: A Typology of POS Taggers

POS taggers can be categorized based on the computational approaches they employ. Each methodology represents a stage in the historical evolution of NLP, from handcrafted knowledge to data-driven learning (Jurafsky & Martin, 2023).

- **Rule-Based POS Tagging:** This is one of the earliest methods, relying on a manually crafted set of linguistic rules. These taggers use a dictionary to find possible tags for a word and then apply disambiguation rules based on the context of surrounding words. While precise and interpretable, rule-based systems are labor-intensive to create and not robust enough to handle novel language patterns (Brill, 1992; Xue & Zhang, 2020; Nguyen et al., 2016).

- **Stochastic (or Probabilistic) POS Tagging:** This approach employs statistical models to assign tags based on probabilities derived from large, annotated corpora. By calculating the likelihood of a tag sequence for a given sentence, these models can effectively handle ambiguity. They are more adaptable than rule-based systems but are highly dependent on the quality and size of their training data (Charniak, 1993).

- **Transformation-Based Tagging:** Also known as the Brill Tagger, this method is a hybrid approach. It begins with an initial tagging (often based on simple probabilities) and then iteratively refines it by applying a learned set of transformation rules that correct common errors based on contextual patterns (Brill, 1995; Nguyen et al., 2016; Lv et al., 2016).

- **Machine Learning-Based POS Tagging:** Modern taggers increasingly rely on machine learning algorithms such as decision trees, support vector machines, and, most prominently, deep learning models. Architectures like recurrent neural networks (RNNs), long short-term memory (LSTM) networks, and transformers (e.g., BERT) have set new standards for accuracy. These models excel at capturing complex, long-range dependencies in text but require substantial annotated data and computational resources for training (Goldberg, 2017; Ullah et al., 2024; Dalai et al., 2023).

6.5 A Deep Dive into Probabilistic Methods: The Hidden Markov Model (HMM)

To understand the foundations of stochastic tagging, we will examine the hidden Markov model (HMM), a classic probabilistic framework for modeling sequential data. HMMs were a dominant technology in NLP for many years and remain an excellent pedagogical tool for understanding probabilistic reasoning (Rabiner, 1989). In an HMM, a system is assumed to transition between hidden states while producing observable outputs. For POS tagging:

- The **hidden states** are the POS tags (noun, verb, etc.), which we cannot directly observe.

- The **observations** are the words in the sentence, which are visible.

The model is defined by three core components (Manning & Schütze, 1999):

1. **States:** A set of hidden states (e.g., all possible POS tags in a tagset).

2. **Observations:** The vocabulary of words that can be emitted by the states.

3. **Probabilities:**

   - **Transition Probabilities:** The likelihood of moving from one state to another. This captures grammatical structure (e.g., the probability that a noun follows a determiner, P(Noun | Determiner)).

   - **Emission Probabilities:** The likelihood of observing a specific word given a particular hidden state. This captures word semantics (e.g., the probability of the word 'run' being a verb, P(run | Verb)).

Using these probabilities, the HMM finds the most probable sequence of tags T for a given sequence of words W by maximizing the joint probability, which simplifies to:

Equation 6.1

Hidden Markov Model (HMM)

$$\arg\max_{T} P(T|W) \propto \prod_{i} P(w_i|t_i) \cdot P(t_i|t_{i-1})$$

where $P(w_i \mid t_i)$ is the emission probability and $P(t_i \mid t_{i-1})$ is the transition probability.

6.6 The Viterbi Algorithm: Decoding the Optimal Tag Sequence

The **Viterbi algorithm** is a dynamic programming method used to efficiently find the most likely sequence of hidden states (POS tags) in an HMM. Instead of calculating the probability of every possible tag sequence (which would be computationally intractable), Viterbi finds this optimal path in polynomial time (Viterbi, 1967).

Let's walk through a conceptual and numerical example for the sentence: **"She runs fast."**

Assume we have the following probabilities derived from a training corpus.

1. Transition Probabilities P(tag₂ | tag₁)

These show the likelihood of one POS tag following another.

Table 6.1

Transition Probabilities

From → To	PRP → VB	PRP → NN	VB → RB	VB → JJ
Probability	0.8	0.2	0.7	0.3

2. Emission Probabilities P(word | tag)

These show the likelihood of a word belonging to a tag.

Table 6.2

Emission Probabilities

Word → POS	P(She | PRP)	P(runs | VB)	P(runs |NN)	P(fast | RB)	P(fast | JJ)
Probability	1.0	0.6	0.4	0.7	0.3

Step-by-Step Viterbi Execution:

1. Initialization (First Word: 'She')
   The word 'She' can only be a Pronoun (PRP).

   - *Path Probability:* P('She' | PRP) = 1.0

   - The algorithm stores this probability and the path: (PRP).

1. Recursion Step 1 (Second Word: 'runs')
   The word 'runs' could be a Verb (VB) or a Noun (NN). The algorithm
   calculates the probability of arriving at each possible state for 'runs' from the

previous state (PRP).

- o *Path to VB:* P(previous_path) * P(VB | PRP) * P('runs' | VB) = 1.0 * 0.8 * 0.6 = 0.48

- o Path to NN: P(previous_path) * P(NN | PRP) * P('runs' | NN) = 1.0 * 0.2 * 0.4 = 0.08
  The algorithm compares these path probabilities. Since 0.48 > 0.08, it prunes the path going to NN and keeps only the most likely path to this point: (PRP → VB) with a cumulative probability of 0.48.

1. Recursion Step 2 (Third Word: 'fast')
   The word 'fast' could be an Adverb (RB) or an Adjective (JJ). The algorithm calculates the probabilities from the end of the previous best path (VB).

   - o *Path to RB:* P(previous_path) * P(RB | VB) * P('fast' | RB) = 0.48 * 0.7 * 0.7 = 0.2352

   - o Path to JJ: P(previous_path) * P(JJ | VB) * P('fast' | JJ) = 0.48 * 0.3 * 0.3 = 0.0432
     The path to RB is more probable (0.2352 > 0.0432).

1. Termination and Path Traceback
   At the end of the sentence, the algorithm identifies the most probable final state (RB) and traces back the path that led to it.

Final POS Tag Sequence (Best Path):

   She (PRP) → runs (VB) → fast (RB)

6.7 Strengths and Limitations of HMMs in the Modern NLP Landscape

Hidden Markov model provided a robust and efficient framework for POS tagging that dominated the field for years. Their strength lies in their simplicity, efficiency with the Viterbi algorithm, and their ability to model local syntactic dependencies effectively (Rabiner, 1989). However, HMMs have significant limitations, primarily stemming from two core assumptions (Manning & Schütze, 1999):

1. **The Markov Assumption:** HMMs assume that the current tag depends only on

the immediately preceding tag. This prevents the model from capturing long-range dependencies, which are common in human language. For example, in "The cats that the dog chased ran away," the verb 'ran' depends on the distant subject 'cats', a relationship an HMM cannot easily model.

2. **The Independence Assumption:** HMMs assume that the current word's tag is determined only by that word itself, ignoring the influence of surrounding words. This "tunnel vision" means that when deciding the tag for an ambiguous word like 'book' in "We should book a flight," the model only considers the probabilities of 'book' being a noun versus a verb (P(book|Noun) vs. P(book|Verb)). It is blind to the powerful contextual clues from the surrounding words 'should' and 'flight,' which clearly indicate a verb is needed.

These limitations, particularly the strict independence assumption, severely restrict the types of information the model can use. A human annotator, when faced with an ambiguous word, would naturally look at a rich set of features: Does the word end in '-in'? Is it capitalized? What was the previous word? What is the next word? An HMM cannot incorporate these rich, overlapping features into its probability calculations. The model's observation at any given time is strictly the word itself. This inability to leverage a broader feature set is a primary reason why HMMs were eventually superseded by discriminative models. These newer models were designed specifically to overcome this limitation by allowing the model to learn from a diverse and arbitrary set of features from the input sequence, leading to a significant improvement in tagging accuracy (Sutton & McCallum, 2012).

6.8 The Rise of Discriminative Models: Conditional Random Fields (CRFs)

The limitations of generative models, such as HMMs, led directly to the development of discriminative models for sequence labeling tasks. Instead of modeling the **joint probability** of tags and words occurring together—denoted by the *comma* in P(texttags , textwords)—a **discriminative model** directly computes the conditional probability of a tag sequence given that a specific word sequence has already been observed, denoted by the *vertical bar* in P(texttags | textwords). Here, 'texttags' represents the entire sequence of output labels (e.g., POS tags) and 'textwords' represents the entire input word sequence (the sentence). This subtle shift in focus is incredibly powerful. It means the model no longer needs to worry about how the underlying words are 'generated'; it can focus all of its modeling capacity on the prediction task itself.

The most successful and influential of these models for POS tagging was the **conditional random field (CRF)**, particularly the linear-chain CRF (Lafferty et al., 2001; Phan et al., 2007; Corro & Roux, 2025). The key advantage of a CRF is its ability to incorporate a rich, arbitrary, and overlapping set of features from the observation sequence into its decision-making process. While an HMM is limited to evaluating the current word in isolation, a CRF utilizes a set of feature functions that describe the relationship between a given tag at a certain position and any aspect of the entire input sentence.

Because the model computes the conditional probability of the sequence, these feature functions can ask complex, simultaneous questions about the text. For instance, a **feature function** could evaluate whether the current word is 'run' while the assigned tag is a verb. Other functions might analyze morphological cues, such as assessing whether a word ending in '-tion' is tagged as a noun, or whether a capitalized word is appropriately assigned the proper noun tag. Furthermore, these functions can capture structural syntactic dependencies, such as verifying if a verb tag immediately follows a noun tag within the predicted sequence. The CRF learns a weight for each of these feature functions during training; a positive weight indicates that the feature is a strong indicator of a correct tag, while a negative weight suggests the opposite.

Table 6.3

Generative versus Discriminative Models

Model Type	Focus of Learning (What is learned?)	Source of Learning (Where is it learned from?)
Generative (HMMs)	The **Joint Probability** of the entire system $P(t, w)$.	The model must implicitly learn the complex relationships between all possible word and tag pairs in the **training set**.
Discriminative (CRFs)	The **Conditional Probability** of the tags, given the words $P(t \mid w)$.	The model learns the optimal weights for a rich set of features **extracted directly from the observation sequence** (the input words, w) in the **labeled training data**.

When tagging a new sentence, the model calculates a score for each possible tag sequence by summing the weights of all active feature functions. Crucially, a CRF normalizes these scores globally across the entire sequence, allowing it to find the single best tag sequence and avoiding the "label bias problem" that plagued earlier discriminative models like maximum entropy Markov models (MEMMs) (McCallum et al., 2000). This ability to learn from a diverse set of handcrafted features made CRFs the state of the art in POS tagging for many years.

6.9 The Modern Standard: The BiLSTM-CRF Architecture

The advent of deep learning brought another paradigm shift. While CRFs were powerful, they relied on extensive and clever manual feature engineering. The goal of modern neural architectures is to learn these features automatically. For sequence labeling tasks like POS tagging, the de facto standard architecture became a hybrid model: the **bidirectional LSTM-CRF (BiLSTM-CRF)** (Huang et al., 2015; Duan et al., 2019). This name stands for a Bidirectional Long Short-Term Memory network combined with a conditional random field layer. This model elegantly combines the strengths of deep representation learning with the structured prediction capabilities of CRFs.

The model has two main components:

1. **The BiLSTM Layer:** The model typically takes as input a sequence of word embeddings. These embeddings are fed into a bidirectional long short-term memory (BiLSTM) network. As discussed in Chapter 2, an LSTM is a type of RNN that is capable of capturing long-range dependencies. By making it bidirectional, the model processes the sentence in both the forward (left-to-right) and backward (right-to-left) directions. For each word, the outputs from the forward and backward LSTMs are concatenated. The result is that the representation for each word is a rich, context-aware vector that incorporates information from every other word in the sentence. In essence, the BiLSTM acts as a powerful, automatic feature extractor, learning the optimal features for the tagging task without any manual engineering (Lample et al., 2016).

Figure 6.2

The Modern POS Tagger: BiLSTM-CRF Architecture

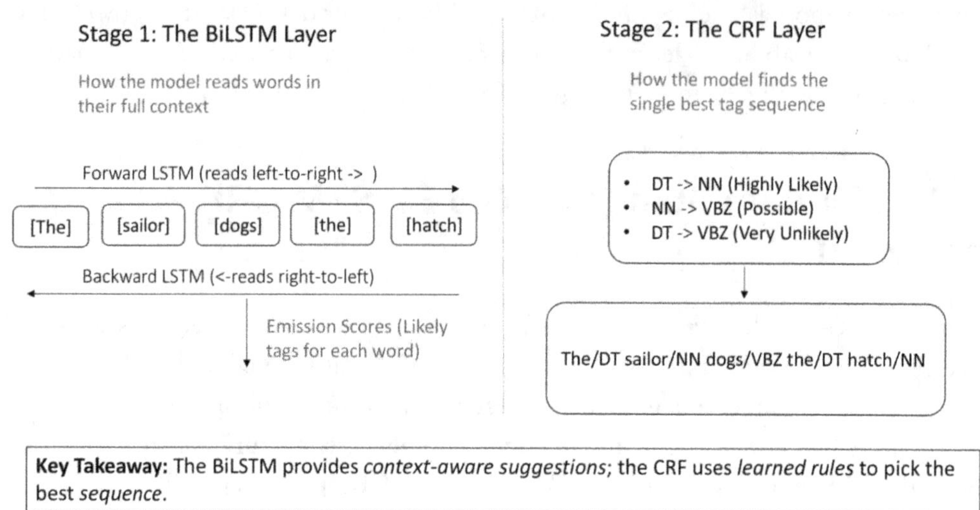

Key Takeaway: The BiLSTM provides *context-aware suggestions*; the CRF uses *learned rules* to pick the best *sequence*.

2. **The CRF Layer:** The output of the BiLSTM layer for each word is a vector of scores, indicating the likelihood of that word belonging to each possible tag. One could simply pick the tag with the highest score at each position (a greedy approach), but this ignores the relationships between adjacent tags. This is where the CRF layer comes in. It sits on top of the BiLSTM and does not predict each tag independently. Instead, it takes the entire sequence of scores from the BiLSTM as input and models the dependencies between tags. The CRF layer learns a matrix of transition scores, representing the likelihood of one tag following another. For example, it will learn that the transition from a *determiner* to a *noun* is highly probable, while the transition from a *determiner* to a *verb* is highly improbable. By combining the BiLSTM's emission scores with its own learned transition scores, the CRF layer can then use the Viterbi algorithm to decode the single most likely tag sequence for the entire sentence, ensuring that the final output is not just a series of locally plausible tags, but a globally coherent and valid sequence (Ma & Hovy, 2016).

6.10 Evaluating POS Taggers

To measure and compare the performance of different POS taggers, a standard set of evaluation practices is used. The primary goal is to assess how well a tagger's output matches a "gold standard" test set manually annotated by human linguists.

The most common metric is **per-word accuracy**, which is simply the percentage of words in the test set that were assigned the correct tag by the model (Manning, 2011). While straightforward, this single number can be misleading. Therefore, a more thorough evaluation involves several key components:

- **Baselines and Upper Bounds:** Performance is never evaluated in a vacuum. A model's accuracy should be compared against a simple **baseline** to demonstrate its value. A common baseline is a unigram tagger that assigns the most frequent tag observed for each word in the training set. The **upper bound** for performance is typically considered to be the **inter-annotator agreement (IAA)** rate. Since human linguists sometimes disagree on the correct tag for ambiguous cases, the IAA (often measured with metrics like Cohen's Kappa) represents the highest achievable performance on the task (Artstein & Poesio, 2008).

- **Confusion Matrix:** A detailed error analysis is crucial for understanding a model's weaknesses. A **confusion matrix** is a table that shows the frequency of each type of error. For example, it might reveal that a model frequently confuses past-tense verbs (VBD) with past participles (VBN), or singular nouns (NN) with plural nouns (NNS). This allows researchers to focus their efforts on improving the model's performance on these specific, challenging cases (Sokolova & Lapalme, 2009).

- **Out-of-Vocabulary (OOV) Performance:** A critical test of a tagger's ability to generalize is its performance on words that were not seen during training. A model's accuracy on OOV words is often reported separately from its overall accuracy. Models that can leverage sub-word information (like character-level embeddings or morphological features) typically perform much better on OOV words than models that treat words as atomic units (Weischedel et al., 2011; Passban et al., 2017).

6.11 Conclusion

Part-of-speech tagging remains a cornerstone of natural language processing, providing the essential grammatical information required to bridge the gap between human language and computational understanding. By resolving the inherent ambiguity of words in context, POS tagging enables more precise and powerful analysis in tasks ranging from machine translation to sentiment analysis.

This chapter has traced the evolution of POS tagging methodologies, from foundational probabilistic models like the hidden Markov model to the state-of-the-art deep learning architectures that power modern libraries like spaCy and Stanza. While classic methods like HMMs introduced a powerful way to reason about linguistic sequences, their inherent limitations—particularly in handling long-range dependencies—paved the way for more sophisticated neural models that capture context more holistically.

The journey from theory to practice underscores a key theme in computer science: the continuous development of new models to overcome the limitations of their predecessors. As NLP continues to advance, future research will focus on improving tagger accuracy for low-resource languages and noisy, informal text. Furthermore, developing computationally efficient models for deployment on edge devices and exploring the application of POS tagging to new frontiers like multimodal language understanding will ensure that this fundamental task remains a relevant and evolving area of study.

6.12 References

Artstein, R., & Poesio, M. (2008). Inter-coder agreement for computational linguistics. *Computational Linguistics, 34*(4), 555–596. https://doi.org/10.1162/coli.07-034-R2

Brill, E. (1992). A simple rule-based part of speech tagger. *Proceedings of the Third Conference on Applied Natural Language Processing,* 152–155. https://doi.org/10.3115/974499.974526

Brill, E. (1995). Transformation-based error-driven learning and natural language processing: A case study in part-of-speech tagging. *Computational Linguistics, 21*(4), 543–565.

Charniak, E. (1993). *Statistical Language Learning.* MIT Press.

Chiche, A., & Yitagesu, B. (2022). Part of speech tagging: A systematic review of deep learning and machine learning approaches. *Journal of Big Data, 9*(1), Article 10. https://doi.org/10.1186/s40537-022-00561-y

Corro, C., Lacroix, M., & Roux, J. L. (2025). *Bregman conditional random fields: Sequence labeling with parallelizable inference algorithms.* https://doi.org/10.48550/arxiv.2506.00732

Dalai, T., Mishra, T. K., & Sa, P. K. (2023). Part-of-speech tagging of Odia language using statistical and deep learning based approaches. *ACM Transactions on Asian and Low-Resource Language Information Processing, 22*(6), Article 167. https://doi.org/10.1145/3588900

Das, S. B., Panda, D., Mishra, T. K., & Patra, B. K. (2025). Statistical machine translation for Indic languages. *Natural Language Processing, 31*(2), 328–345. https://doi.org/10.1017/nlp.2024.26

Duan, J., Wang, B., Tan, Z., Wei, X., & Wang, H. (2019). Chinese spelling check via bidirectional LSTM-CRF. *2019 IEEE 8th Joint International Information Technology and Artificial Intelligence Conference (ITAIC)*, 1333–1336. https://doi.org/10.1109/ITAIC.2019.8785520

Dubey, S. (2024). Clustering for clarity: Improving word sense disambiguation through multilevel analysis. *Computer Science, 25*(2). https://doi.org/10.7494/csci.2024.25.2.5844

Gholami-Dastgerdi, P., & Feizi-Derakhshi, M.-R. (2023). Part of speech tagging using part of speech sequence graph. *Annals of Data Science, 10*(5), 1301–1328. https://doi.org/10.1007/s40745-021-00359-4

Goldberg, Y. (2017). *Neural network methods for natural language processing.* In Synthesis Lectures on Human Language Technologies (1st ed., Vol. 37). Springer Nature. https://doi.org/10.1007/978-3-031-02165-7

Grishman, R. (2019). Twenty-five years of information extraction. *Natural Language Engineering, 25*(6), 677–692. https://doi.org/10.1017/S1351324919000512

Heidorn, G. E. (2000). Intelligent writing assistance. In R. Dale, H. Moisl, & H. Somers (Eds.), *Handbook of natural language processing* (pp. 597–619). Marcel Dekker.

Hobbs, J. R. (1978). Resolving pronoun references. *Lingua, 44*(4), 311–338. https://doi.org/10.1016/0024-3841(78)90006-2

Huang, Z., Xu, W., & Yu, K. (2015). *Bidirectional LSTM-CRF models for sequence tagging.* arXiv preprint arXiv:1508.01991. https://doi.org/10.48550/arXiv.1508.01991

Jurafsky, D., & Martin, J. H. (2023). *Speech and language processing* (3rd ed.). Prentice Hall.

Kaur, N., & Singh, P. (2023). Conventional and contemporary approaches used in text to speech synthesis: a review. *The Artificial Intelligence Review, 56*(7), 5837–5880. https://doi.org/10.1007/s10462-022-10315-0

Keith Norambuena, B., Lettura, E. F., & Villegas, C. M. (2019). Sentiment analysis and opinion mining applied to scientific paper reviews. *Intelligent Data Analysis, 23*(1), 191–214. https://doi.org/10.3233/IDA-173807

Koehn, P. (2009). *Statistical machine translation* (Reprint. with corrections). Cambridge University Press. https://doi.org/10.1017/CBO9780511815829

Lafferty, J., McCallum, A., & Pereira, F. C. (2001). Conditional random fields: Probabilistic models for segmenting and labeling sequence data. *Proceedings of the Eighteenth International Conference on Machine Learning*, 282–289.

Lample, G., Ballesteros, M., Subramanian, S., Kawakami, K., & Dyer, C. (2016). *Neural architectures for named entity recognition.* arXiv preprint arXiv:1603.01360. https://doi.org/10.48550/arXiv.1603.01360

Liu, R., Mao, R., Luu, A. T., & Cambria, E. (2023). A brief survey on recent advances in coreference resolution. *The Artificial Intelligence Review, 56*(12), 14439–14481. https://doi.org/10.1007/s10462-023-10506-3

Lv, C., Liu, H., Dong, Y., & Chen, Y. (2016). Corpus based part-of-speech tagging. International *Journal of Speech Technology, 19*(3), 647–654. https://doi.org/10.1007/s10772-016-9356-2

Ma, X., & Hovy, E. (2016). *End-to-end sequence labeling via bi-directional LSTM-CNNs-CRF.* arXiv preprint arXiv:1603.01354. https://doi.org/10.48550/arXiv.1603.01354

Manning, C. D. (2011). Part-of-speech tagging from 97% to 100%: Is it time for some linguistics? In A. F. Gelbukh (Ed.), *Computational Linguistics and Intelligent Text Processing* (pp. 171–189). Springer Berlin Heidelberg. https://doi.org/10.1007/978-3-642-19400-9_14

Manning, C. D., & Schütze, H. (1999). *Foundations of statistical natural language processing.* MIT Press.

McCallum, A., Freitag, D., & Pereira, F. (2000). Maximum entropy Markov models for information extraction and segmentation. *Proceedings of the Seventeenth International Conference on Machine Learning,* 591–598.

Moldovan, D., Harabagiu, S., Pasca, M., Mihalcea, R., Girju, R., Goodrum, R., & Rus, V. (2003). LCC: The first question answering system. In M. T. Maybury (Ed.), *New directions in question answering* (pp. 11–20). AAAI Press.

Nadeau, D., & Sekine, S. (2007). A survey of named entity recognition and classification. *Lingvisticae Investigationes, 30*(1), 3–26. https://doi.org/10.1075/li.30.1.03nad

Navigli, R. (2009). Word sense disambiguation: A survey. *ACM Computing Surveys, 41*(2), 1–69. https://doi.org/10.1145/1459352.1459355

Nenkova, A., & McKeown, K. (2012). *A survey of text summarization techniques.* In C. C. Aggarwal & C. Zhai (Eds.), Mining text data (pp. 43–76). Springer.

Nguyen, D. Q., Nguyen, D. Q., Pham, D. D., & Pham, S. B. (2016). A robust transformation-based learning approach using ripple down rules for part-of-speech tagging. *AI Communications,* 29(3), 409–422. https://doi.org/10.3233/aic-150698

Ojokoh, B., & Adebisi, E. (2019). A review of question answering systems. *Journal of Web Engineering, 17*(8), 717–758.

Pang, B., & Lee, L. (2008). Opinion mining and sentiment analysis. *Foundations and Trends in Information Retrieval, 2*(1–2), 1–135. https://doi.org/10.1561/1500000011

Passban, P., Liu, Q., & Way, A. (2017). Boosting neural POS tagger for Farsi using morphological information. *ACM Transactions on Asian and Low-Resource Language Information Processing, 16*(1), 1–15. https://doi.org/10.1145/2934676

Rabiner, L. R. (1989). A tutorial on hidden Markov models and selected applications in speech recognition. *Proceedings of the IEEE, 77*(2), 257–286. https://doi.org/10.1109/5.18626

Phan, X.-H., Nguyen, L.-M., Inoguchi, Y., & Horiguchi, S. (2007). High-performance training of conditional random fields for large-scale applications of labeling sequence data. *IEICE Transactions on Information and Systems*, E90-D(1), 13–21. https://doi.org/10.1093/ietisy/e90-1.1.13

Reddy, B. V., Rao, K. S., & Neeraja, K. (2023). Named entity recognition on different languages: A survey. In M. V. Reddy, M. S. Gupta, & A. V. Anand (Eds.), *AIP Conference Proceedings* (Vol. 2492,

Number 1). American Institute of Physics. https://doi.org/10.1063/5.0113210

Sharma, G., & Sharma, D. (2023). Automatic text summarization methods: A comprehensive review. *SN Computer Science, 4*(1), Article 33. https://doi.org/10.1007/s42979-022-01446-w

Sokolova, M., & Lapalme, G. (2009). A systematic analysis of performance measures for classification tasks. *Information Processing & Management, 45*(4), 427–437. https://doi.org/10.1016/j.ipm.2009.03.002

Sutton, C., & McCallum, A. (2012). An introduction to conditional random fields. *Foundations and Trends in Machine Learning, 4*(4), 267–373. https://doi.org/10.1561/2200000013

Taylor, P. (2009). *Text-to-speech synthesis.* Cambridge University Press. https://doi.org/10.1017/cbo9780511816338

Ullah, S., Ahmad, R., Namoun, A., Muhammad, S., Ullah, K., Hussain, I., & Ibrahim, I. A. (2024). A deep learning-based approach for part of speech (POS) tagging in the Pashto language. *IEEE Access, 12,* 86355–86364. https://doi.org/10.1109/ACCESS.2024.3412175

Viterbi, A. J. (1967). Error bounds for convolutional codes and an asymptotically optimum decoding algorithm. *IEEE Transactions on Information Theory, 13*(2), 260–269. https://doi.org/10.1109/TIT.1967.1054010

Weischedel, R., Licuanan, A., & Ramshaw, L. (2011). The DARPA machine reading program—encouraging multilingual and deep natural language understanding. *Proceedings of the 49th Annual Meeting of the Association for Computational Linguistics: Human Language Technologies,* 1–6.

Xue, X., & Zhang, J. (2020). Building codes part-of-speech tagging performance improvement by error-driven transformational rules. *Journal of Computing in Civil Engineering, 34*(5). https://doi.org/10.1061/(ASCE)CP.1943-5487.0000917

6.13 Glossary

Annotated Corpora: Large collections of text that have been manually or semi-automatically tagged with linguistic information, such as part-of-speech tags, used for training and evaluating NLP models.

BiLSTM-CRF: A state-of-the-art hybrid neural architecture for sequence labeling that combines a Bidirectional LSTM to learn long-range, context-aware features and a CRF layer to ensure the final tag sequence is globally optimal.

Conditional Random Field (CRF): A discriminative statistical model used for sequence labeling that models the conditional probability of a tag sequence given an input sequence, allowing it to incorporate a rich set of overlapping features.

Confusion Matrix: A table used in evaluation that visualizes the performance of a classification model by showing the counts of correct and incorrect predictions for each class.

Constituency Parsing: An NLP task that breaks down a sentence into its constituent phrases (like noun phrases, verb phrases), representing the grammatical structure in a tree format.

Dependency Parsing: An NLP task that analyzes the grammatical relationships between words in a sentence, typically represented as a tree structure showing which words modify or depend on others.

Discriminative Model: A type of statistical model that learns the conditional probability of an output given an input ($P(Y|X)$), focusing directly on the prediction task.

Emission Probabilities (in HMM): The likelihood of observing a specific output given a particular hidden state in an HMM (e.g., the probability of seeing the word 'bank' when the state is 'noun').

Generative Model: A type of statistical model that learns the joint probability of inputs and outputs ($P(X,Y)$), allowing it to 'generate' new data points.

Hidden Markov Model (HMM): A generative probabilistic model that models sequential data by assuming a system transitions between hidden states while producing observable outputs.

Inter-annotator Agreement (IAA): A measure of how well two or more human annotators agree when labeling the same data, often used as an upper bound for model performance.

Machine Translation: The task of automatically translating text or speech from one language to another.

Named Entity Recognition (NER): An NLP task that involves identifying and classifying named entities in text, such as names of people, organizations, locations, etc.

Natural Language Processing (NLP): A field of computer science and artificial intelligence concerned with the interactions between computers and human language.

Observations (in HMM): The visible outputs generated by the hidden states in an HMM, such as words in a sentence.

Part-of-Speech (POS) Tagging: The process of assigning grammatical categories (like noun, verb, adjective) to words in a text based on their definition and context.

Part-of-Speech Tagger: A computational tool or algorithm used to perform Part-of-Speech tagging.

Per-word Accuracy: The standard evaluation metric for POS tagging, calculated as the percentage of words in a test set that were assigned the correct tag.

Rule-Based POS Tagging: A method that uses predefined linguistic rules to assign POS tags.

Sentiment Analysis: The process of computationally identifying and categorizing opinions expressed in a piece of text, typically determining whether the attitude is positive, negative, or neutral.

SpaCy: A fast and efficient Python library for NLP, designed for production use and leveraging deep learning models.

Stanza: A suite of NLP tools from Stanford University, accessible in Python, known for its accuracy and multilingual support.

States (in HMM): Hidden variables or underlying conditions within an HMM, representing concepts like POS tags in the context of language.

Stochastic POS Tagging: A method that uses statistical models and probabilities derived from training data to assign POS tags.

Transition Probabilities (in HMM): The likelihood of moving from one hidden state to another in an HMM (e.g., the probability of transitioning from a determiner tag to a noun tag).

Viterbi Algorithm: A dynamic programming algorithm used to find the most probable sequence of hidden states in a hidden Markov model, commonly used for decoding the best tag sequence in HMM-based POS tagging.

Word Sense Disambiguation (WSD): The computational task of determining the intended meaning of a word in a specific context when that word has multiple possible meanings.

6.14 Review Questions and Discussion Topics

1. What is the primary goal of part-of-speech (POS) tagging?

2. Explain the main difference between a part-of-speech tagger and part-of-speech tagging.

3. Name two applications of POS tagging in natural language processing (NLP).

4. In the context of HMM-based POS tagging, define "emission probabilities."

5. What is the main function of the Viterbi algorithm in HMM-based POS tagging?

6. What is a key advantage of a Conditional Random Field (CRF) over a Hidden Markov Model (HMM)?

7. Describe the two main components of the modern BiLSTM-CRF architecture.

8. What is per-word accuracy, and why is it important to also consider a model's performance on out-of-vocabulary (OOV) words?

9. Compare and contrast a generative model like an HMM with a discriminative model like a CRF for the task of POS tagging. Why was the shift to discriminative models considered a major advance in the field?

10. Explain the roles of the BiLSTM layer and the CRF layer in the BiLSTM-CRF architecture. Why is this hybrid model generally more effective than using a BiLSTM with an output layer that predicts each tag independently?

11. Imagine you are tasked with evaluating a new POS tagger for English. Describe the steps you would take to perform a rigorous evaluation, including the metrics, baselines, and types of analysis you would use.

12. Discuss the significance of part-of-speech tagging as a preprocessing step in natural language processing. Choose three different NLP applications mentioned in the text and explain in detail how POS tagging contributes to their functionality.

Chapter 7

The Penn Treebank Tagset: A Foundation for Computational Linguistics

7.1 Introduction: Giving Language a Grammatical Blueprint

Before a machine can understand the meaning of a sentence, it must first understand the role each word plays—is it a noun, a verb, an adjective? This fundamental task, known as part-of-speech (POS) tagging, is a cornerstone of natural language processing (NLP). In the late 1980s, the field needed a standardized, computationally efficient blueprint for English grammar. The Penn Treebank project met this need by creating a large, meticulously annotated corpus of English text, providing researchers with a shared resource for both syntactic structure and part-of-speech information.

The result was the Penn Treebank tagset, a standardized set of POS tags that became a de facto standard in computational linguistics. Its design principles struck a critical balance between linguistic detail and computational pragmatism, simplifying earlier, more complex tagsets for greater efficiency. While modern NLP has evolved to include complex neural models, the legacy of the Penn Treebank endures. Many advanced models are still trained and evaluated on datasets annotated with their tags, and this influence is visible in the architectures of tools for parsing, machine translation, and information extraction. We will examine its design philosophy, its core applications, and provide concrete examples of its use. Furthermore, we will analyze how its design compares to other major tagsets and conclude with a discussion of its lasting legacy and the future challenges in syntactic annotation.

7.2 The Penn Treebank Tagset: Design and Philosophy

The Penn Treebank tagset was designed with a philosophy of "computational pragmatism." The creators aimed for a set of tags that was detailed enough to be useful for syntactic parsing but simple enough to allow human annotators to apply them consistently and for computational models to learn them effectively. This led to several key design principles (Marcus et al., 1993):

- **Simplicity and Consistency:** The tagset is relatively small, containing 36 POS tags and 12 tags for punctuation and currency symbols. This contrasts with earlier, more complex tagsets like the Brown Corpus, which had over 80 tags.

- **Information Recovery:** The tagset omits distinctions that are "lexically recoverable"—that is, information that can be easily determined from the word itself using a dictionary. For example, it does not distinguish between the different forms of the verb "to be" (am, is, are, was, were), as these are all easily identifiable.

- **Encoding of Syntactic Function:** The core innovation of the **Penn Treebank** is its ability to encode the word's syntactic *role* in the sentence (which is necessary for **parsing**) into its tag, whereas the **Brown Corpus** tags are based on the word's *form* or *spelling*.

Table 7.1

Penn Treebank versus Brown Corpus

Sentence Structure	Word	Brown Corpus Tag (Morphological/ Context)	Penn Treebank Tag (Syntactic/ Structural)	Structural Rationale (Penn Treebank)
"He enjoyed *being* polite"	*being*	*BEG* (Gerund/ Nominal Verb)	*VBG* (Verb, Gerund/Pres. Participle)	The Penn Treebank uses *VBG* to signal the word's role as a noun/gerund object of the main verb ('enjoy').

7.2.1 Design in Practice: The Principle of Lexical Recoverability

To fully appreciate the "computational pragmatism" of the Penn Treebank, it is useful to examine a specific design choice through the lens of lexical recoverability. A key decision was not to always distinguish between the past tense (VBD) and past participle (VBN) forms of verbs. For regular English verbs, these two forms are identical (e.g., 'walked,' 'finished'). For many irregular verbs, however, they are distinct (e.g., 'ate' vs. 'eaten'; 'drove' vs. 'driven') (Santorini, 2023).

The Penn Treebank's philosophy is to only create a separate tag if the distinction cannot be easily recovered from the word itself. Consider the verb 'drive.' The past tense form is 'drove,' and the past participle is 'driven.' Because these two words are unique, a downstream application with access to a lexicon can unambiguously identify 'driven' as a past participle. Therefore, the tagset designers argued that creating a separate tag for it was redundant. However, for a word like 'finish,' the form 'finished' could be either the past tense (as in "He finished the race") or a past participle (as in "He has finished the race"). Because the word form itself is ambiguous, a distinct tag is necessary to resolve that ambiguity for a parser. This principle allows the tagset to remain small and focused, offloading unambiguous distinctions to a dictionary and reserving the tags for resolving genuine structural ambiguities that are critical for syntactic analysis. This trade-off—a smaller, more manageable tagset at the cost of requiring a lexicon for full analysis—is a hallmark of the Penn Treebank's pragmatic design (Marcus et al., 1993).

7.2.2 A Reference Guide to Common Penn Treebank Tags

To serve as a practical reference, the following table details the most frequently used tags in the Penn Treebank set, along with their descriptions and clear examples. Understanding these core tags is essential for anyone working with NLP tools and corpora that use this standard (Santorini, 2023).

Table 7.2

Core Tags in The Penn Treebank Set

Tag	Description	Example
NN	Noun, singular or mass	"The *cat* sat on the *mat*"
NNS	Noun, plural	"The *cats* chased the mice"
NNP	Proper noun, singular	"*Google* is a large *company*"
PRP	Personal pronoun	"*He* saw *her* at the store"
DT	Determiner	"*The* cat, *a* friend, *an* apple"
JJ	Adjective	"The *quick, brown* fox"
RB	Adverb	"He ran *quickly*"
IN	Preposition / Subordinating Conj.	"*In* the house; *because* it rained"
CC	Coordinating conjunction	"bread *and* butter"
VB	Verb, base form	"I want to *run*"
VBP	Verb, non-3rd person singular present	"I *run*; they *run*"
VBZ	Verb, 3rd person singular present	"He *runs*"
VBD	Verb, past tense	"She *ran* yesterday"
VBG	Verb, gerund or present participle	"*Running* is fun"
VBN	Verb, past participle	"He has *run* a marathon"
TO	to	"I want *to* go"
CD	Cardinal number	"There are *three* cats"

7.3 Applications in Natural Language Processing

The consistency and standardization of the Penn Treebank tagset make it an invaluable tool for a wide range of NLP applications. By providing a machine-readable grammatical structure, the tags serve as the foundation for more complex language understanding tasks (Jurafsky & Martin, 2023).

- **Syntactic Parsing:** Parsing aims to determine the grammatical structure of a sentence. The POS tags from the Penn Treebank are essential for this, as they provide the lexical categories that a parser uses to build a full syntactic tree.

- **Named Entity Recognition (NER):** NER systems rely on POS tags to help identify proper nouns (NNP), which are strong indicators of named entities like people, organizations, and locations (Nadeau & Sekine, 2007).

- **Information Extraction:** By identifying nouns, verbs, and adjectives, POS tagging helps systems extract key information and relationships from text.

- **Sentiment Analysis:** POS tags can help sentiment analysis models identify sentiment-bearing words, such as adjectives (JJ) and adverbs (RB), to determine the emotional tone of a text (Pang & Lee, 2008).

- **Machine Translation:** Corpora annotated with Penn Treebank tags are frequently used to train machine translation models, helping them generate grammatically correct translations in the target language (Koehn, 2009).

7.3.1 The Primary Goal: Enabling Syntactic Parsing

While POS tagging is a useful task in its own right, the primary motivation for the Penn Treebank project was to facilitate the much larger goal of full **syntactic parsing**. The POS tags serve as the terminal nodes, or 'leaves,' in a complete constituency parse tree. The project also defined a set of **phrase-level tags** to label the internal nodes of the tree, creating a complete grammatical blueprint of a sentence (Marcus et al., 1993).

The most common phrase-level tags include:

- **S:** Sentence (the root of the tree)

- **NP:** Noun Phrase (e.g., "the quick brown fox")

- **VP:** Verb Phrase (e.g., "jumps over the lazy dog")

- **PP:** Prepositional Phrase (e.g., "over the lazy dog")

Using both the POS tags and these phrase-level tags, our example sentence can be represented in a bracketed format that shows its complete hierarchical structure:

(S (NP (DT The) (JJ quick) (JJ brown) (NN fox)) (VP (VBZ jumps) (PP (IN over) (NP (DT the) (JJ lazy) (NN dog)))))

This representation makes the grammatical structure explicit. It clearly shows that "The quick brown fox" is a single noun phrase acting as the subject, and "jumps over the lazy dog" is a verb phrase. Within that VP, "over the lazy dog" is a prepositional phrase that modifies the verb "jumps." A computational parser's job is to take a sentence as input and produce this structured output. The POS tags are the essential first step in this process, providing the parser with the foundational word-level categories it needs to begin building the larger phrasal structures (Manning & Schütze, 1999).

Figure 7.1

The Full Constituency Parse Tree

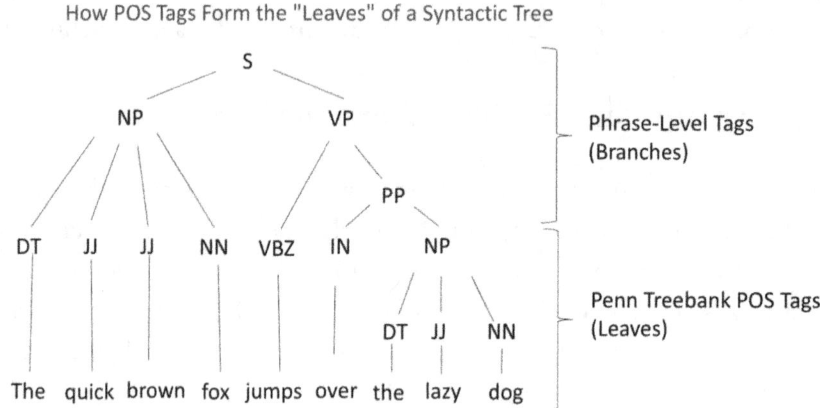

7.4 The Challenge of Ambiguity

A significant challenge in POS tagging is ambiguity: a single word can have different parts of speech depending on context. The Penn Treebank system acknowledges this by allowing words to be associated with multiple tags when the context is insufficient for a definitive choice, although the final corpus provides a single disambiguated tag for each word (Marcus et al., 1993). Consider the sentence: "They are flying planes." This sentence is ambiguous. 'Flying' could be part of a verb phrase ("they are currently flying") or it could be an adjective describing the planes. A tagger must use context to disambiguate:

1. They/PRP are/VBP flying/VBG planes/NNS. (Here, 'flying' is a present participle, part of the main verb.)

2. They/PRP are/VBP flying/JJ planes/NNS. (Here, 'flying' is an adjective modifying 'planes'.)

7.4.1 A Case Study in Ambiguity: The IN Tag

One of the most debated design choices in the Penn Treebank is the use of a single tag, IN, to cover both **prepositions** and **subordinating conjunctions**. From a linguistic perspective, these are distinct grammatical classes. Prepositions (e.g., 'in,' 'on,' 'with') introduce prepositional phrases that typically function as adverbs or adjectives. Subordinating conjunctions (e.g., 'because,' 'while,' 'if') introduce dependent clauses that connect to a main clause (Huddleston & Pullum, 2002).

Consider these two sentences:

1. He sat **in/IN** the chair. (Preposition)

2. He sat down **because/IN** he was tired. (Subordinating Conjunction)

Figure 7.2

A Case Study in Ambiguity: The 'IN' Tag

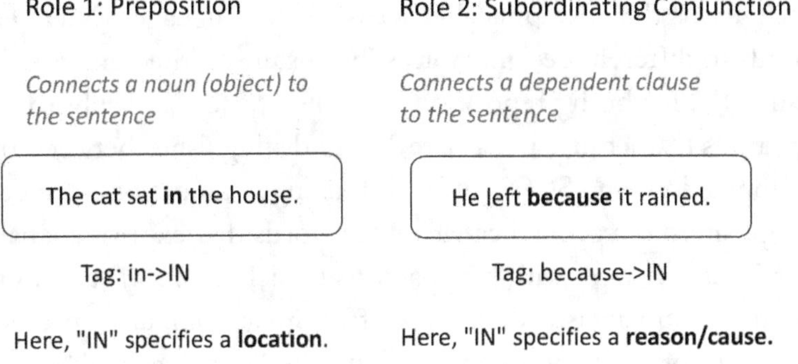

The Penn Treebank tagset conflates these two functions under a single tag. This decision was made for pragmatic reasons: in many cases, the distinction is difficult for human annotators to make consistently, and for many parsing tasks, the distinction was not considered critical. However, this simplification poses a significant challenge for downstream NLP applications that require a deeper semantic understanding. For example, a question-answering system might need to distinguish between a phrase that describes a location (introduced by a preposition) and a clause that describes a reason (introduced by a subordinating conjunction). This ambiguity is a primary reason why modern, cross-linguistically focused annotation schemes like Universal Dependencies (UD) explicitly separate these categories, using adposition (ADP) for prepositions and subordinating conjunction (SCONJ) for subordinating conjunctions. This highlights a key tension in tagset design: the balance between simplicity for annotation and the level of detail required for sophisticated language understanding (de Marneffe et al., 2021).

7.5 Comparison with Other Tagsets

The Penn Treebank's design is best understood in comparison to other tagsets.

- **Brown Corpus Tagset:** The Brown Corpus tagset is an older, more complex system with over 80 tags. It is highly specific, often tied to particular lexical items. The Penn Treebank deliberately simplified this approach to improve computational efficiency and ease of use, arguing that context could recover much of the information lost in simplification (Francis & Kučera, 1979; Marcus et al., 1993).

- **Universal Dependencies (UD):** UD is a modern framework designed for cross-linguistically consistent annotation. Its goal is to create grammatical representations that are applicable across many languages (Nivre et al., 2016). This leads to different design choices. For example, the Penn Treebank uses a single tag, IN, for both prepositions ("in the house") and subordinating conjunctions ("...because it was late"). UD distinguishes between these, using ADP (adposition) and SCONJ (subordinating conjunction), respectively. The term adposition is a general category for words that express spatial or temporal relationships, with prepositions (which come *before* a noun) being a common type in English. While UD offers greater linguistic precision and is better suited for multilingual applications, the Penn Treebank's simplicity and the vast amount of legacy data annotated with its tags ensure its continued relevance, especially for English-specific tasks.

7.6 Legacy and Future Directions

The Penn Treebank tagset has had a profound and lasting impact on the field of NLP. It provided a standardized, large-scale resource that fueled decades of research in statistical parsing and tagging. However, as the field evolves, several challenges and future directions have emerged:

- **Mapping and Interoperability:** As modern frameworks like Universal

Dependencies become more prominent, a key area of research is improving interoperability—the ability for different systems to work together. This involves developing robust, automated tools to map or convert the vast amount of legacy data—valuable information annotated with older standards such as the Penn Treebank tagset—into these new formats. Resolving the differences between these systems is crucial, as it allows the decades of work invested in creating Penn Treebank data to be used effectively with modern NLP tools (de Marneffe et al., 2021).

- **Domain-Specific Adaptation**: The Penn Treebank is primarily based on edited, formal text (like news articles). There is a growing need for models that can adapt to specialized domains, such as legal or biomedical texts. These fields have unique terminologies and syntactic nuances that general models struggle with. The modern approach is often to fine-tune large, pre-trained models on smaller, domain-specific annotated corpora to adapt them to these specialized language patterns (Chalkidis et al., 2020).

- **Advanced Ambiguity Resolution:** Developing more sophisticated methods for resolving tagging ambiguities by integrating richer contextual data remains a valuable path for research, especially for noisy user-generated content like social media text (Gimpel et al., 2011). Recent work continues to demonstrate the effectiveness of domain-specific pretrained language models for handling the unique challenges of social media text processing (Guo & Sarker, 2023), while advances in socially-enriched models trained on billions of tweets show improved performance on short, noisy, user-generated text (Zhang et al., 2023).

As NLP moves further into the era of large language models, the role of explicit, human-designed grammatical annotation, such as the Penn Treebank tagset, is changing. Yet, its core principles—the need for standardized data, the balance between linguistic detail and computational feasibility, and the foundational role of syntax in language understanding—remain as relevant as ever.

7.7 References

Chalkidis, I., Fergadiotis, M., Malakasiotis, P., Aletras, N., & Androutsopoulos, I. (2020). LEGAL-BERT: The Muppets straight out of Law School. *Findings of the Association for Computational Linguistics: EMNLP 2020*, 2898-2904. https://doi.org/10.18653/v1/2020.findings-emnlp.261

de Marneffe, M. C., Manning, C. D., Nivre, J., & Zeman, D. (2021). *Universal Dependencies. Computational Linguistics, 47*(2), 255-308. https://doi.org/10.1162/coli_a_00402

Francis, W. N., & Kučera, H. (1979). *Brown Corpus manual: Manual of information to accompany a standard corpus of present-day edited American English, for use with digital computers.* Brown University, Department of Linguistics.

Gimpel, K., Schneider, N., O'Connor, B., Das, D., Mills, D., Eisenstein, J., ... & Smith, N. A. (2011). Part-of-speech tagging for Twitter: Annotation, features, and experiments. *Proceedings of the 49th Annual Meeting of the Association for Computational Linguistics: Human Language Technologies,* 42–47.

Guo, Y., & Sarker, A. (2023). SocBERT: A pretrained model for social media text. *In Proceedings of the Fourth Workshop on Insights from Negative Results in NLP* (pp. 45-52). Association for Computational Linguistics. https://doi.org/10.18653/v1/2023.insights-1.5

Huddleston, R., & Pullum, G. K. (2002). *The Cambridge grammar of the English language.* Cambridge University Press.

Jurafsky, D., & Martin, J. H. (2023). *Speech and language processing* (3rd ed.). Pearson.

Koehn, P. (2009). *Statistical machine translation* (Reprint. with corrections). Cambridge University Press. https://doi.org/10.1017/CBO9780511815829

Manning, C. D., & Schütze, H. (1999). *Foundations of statistical natural language processing.* MIT Press.

Marcus, M. P., Santorini, B., & Marcinkiewicz, M. A. (1993). Building a large annotated corpus of English: The Penn Treebank. *Computational Linguistics, 19*(2), 313–330.

Nadeau, D., & Sekine, S. (2007). A survey of named entity recognition and classification. *Lingvisticae Investigationes, 30*(1), 3–26. https://doi.org/10.1075/li.30.1.03nad

Nivre, J., de Marneffe, M. C., Ginter, F., Goldberg, Y., Hajič, J., Manning, C. D., ... & Zeman, D. (2016). Universal Dependencies v1: A multilingual treebank collection. *Proceedings of the Tenth International Conference on Language Resources and Evaluation (LREC'16),* 1659–1666.

Pang, B., & Lee, L. (2008). Opinion mining and sentiment analysis. *Foundations and trends in Information Retrieval, 2*(1–2), 1–135. https://doi.org/10.1561/1500000011

Santorini, B. (2023). *Part-of-speech tagging guidelines for the Penn Treebank Project* (3rd Revision).

Zhang, X., Malkov, Y., Florez, O., Park, S., McWilliams, B., Han, J., El-Kishky, A., Gupta, A., Guo, S., Al-Onaizan, Y., & Ippolito, D. (2023). TwHIN-BERT: A socially-enriched pre-trained language model for multilingual tweet representations at Twitter. *In Proceedings of the 29th ACM SIGKDD Conference on Knowledge Discovery and Data Mining* (pp. 3360–3373). https://doi.org/10.1145/3580305.3599921

7.8 Glossary

Brown Corpus Tagset: An earlier and more complex set of POS tags used for annotating the Brown Corpus of American English.

Constituency Parsing: An NLP task that breaks down a sentence into its constituent phrases (like noun phrases, verb phrases), representing the grammatical structure in a hierarchical tree format.

Corpus (plural: Corpora): A large and structured set of texts used for linguistic analysis.

Dependency-Based Formats: Linguistic annotation formats that represent grammatical relationships as dependencies between words.

Information Extraction: The task of automatically extracting structured information from unstructured text.

Lexical Recoverability: A design principle in tagsets where linguistic distinctions that can be easily determined from a word's form (using a dictionary) are omitted from the tagset to maintain simplicity.

Machine Translation: The task of automatically translating text or speech from one language to another.

Named Entity Recognition: The task of identifying and classifying named entities in text, such as person names, organizations, locations, and dates.

Natural Language Processing (NLP): A field of computer science that focuses on enabling computers to understand, interpret, and generate human language.

Part-of-Speech (POS) Tagging: The process of labeling each word in a text with its corresponding grammatical category.

Penn Treebank: A large annotated corpus of English text that is a standard resource in NLP.

Sentiment Analysis: The process of determining the emotional tone of a piece of text, such as whether it is positive, negative, or neutral.

Speech Recognition: The process of converting spoken language into text.

Syntactic Parsing: The process of analyzing the grammatical structure of sentences to determine the relationships between words.

Tagging Ambiguity: When a word can be assigned multiple possible part-of-speech tags.

Universal Dependencies (UD): A framework for cross-linguistically consistent grammatical annotation, often presented in a dependency format.

7.9 Review Questions and Discussion Topics

1. **Tagging Ambiguous Sentences:** Provide the Penn Treebank POS tags for the word 'watch' in the following two sentences. Justify your choice of tags based on the word's syntactic function in each sentence.

   A) "Their watch is broken."

   B) "They watch the game from the sidelines."

2. **Manual Annotation:** Select a paragraph from a recent news article. Manually apply the Penn Treebank tags to each word. Note any words where the choice of tag was difficult or ambiguous and explain your reasoning.

3. **Comparative Analysis:** The word 'that' can function as a determiner (DT), a subordinating conjunction (IN), or a pronoun. Write a unique sentence for each use case and apply the correct Penn Treebank tag.

4. Discuss the significance of the Penn Treebank tagset's balance between simplicity and coverage for its effectiveness in computational language analysis.

5. Analyze how the Penn Treebank tagset facilitates key downstream applications in NLP, such as machine translation and semantic analysis, using specific examples from the text.

6. Compare and contrast the design principles and features of the Penn Treebank tagset with Universal Dependencies, highlighting the advantages and disadvantages of each approach.

7. Evaluate the future directions for research and development of the Penn Treebank tagset as outlined in the text, considering their potential impact on advancing NLP capabilities in the age of large language models.

PART III

The Deep Learning Revolution

Chapter 8

Foundations in Classic Machine Learning for NLP

8.1 Introduction: The Statistical Revolution

Before the rise of deep neural networks, the field of natural language processing was defined by a different kind of revolution: the shift from handcrafted rule-based systems to **statistical machine learning**. From the 1990s through the early 2010s, this paradigm dominated NLP research and applications. Instead of relying on linguists to manually codify language rules, the new approach was to learn patterns directly from data, making NLP systems more **robust, scalable, and adaptable**.

This chapter provides a focused tour of these foundational machine learning models. To maintain a clear distinction, the chapter is conceptually divided into two critical phases. The first phase is **feature engineering** (Section 8.2), which involves the necessary process of converting raw, unstructured text into the numerical feature matrix required by all algorithms. The second phase, **classic classification algorithms** (Section 8.3), covers the probabilistic and linear models that ingest this feature matrix to perform document-level text classification.

Understanding these techniques is not merely a historical exercise; they remain powerful, efficient, and widely used for many text analysis tasks, often serving as **strong baselines** against which more complex deep learning models are measured. We will delve into the mechanics of these classic algorithms and see how their shortcomings—particularly their inability to capture semantics—directly motivated the search for the neural architectures that define the modern era (Bengio et al., 2003).

8.2 Feature Engineering for Text: The Art of Representation

Unlike modern deep learning models, which learn optimal **word embeddings** from the raw numerical representations of data to capture semantic and contextual information, classic machine learning algorithms require text to be explicitly organized as a **numerical feature matrix**. The necessary process of converting raw, unstructured text into a set of informative numerical features is known as **feature engineering**. The quality of these handcrafted features directly determines the model's performance, typically by creating a **vector representation** for each document within a **vector space model,** a foundational concept introduced in the early years of information retrieval (Salton et al., 1975; Wong & Ziarko, 1985).

8.2.0 Defining the Feature in Text Analysis

The term **feature** is central to the classic machine learning paradigm, where it holds a precise technical meaning that differs from its everyday usage. In this context, a feature is defined as a quantifiable, individual input variable used by a model to make a prediction. Critically, a model does not directly read text; it learns by assigning numerical weights to each input variable. The collection of all these features for every document constitutes the **feature matrix**, where each document is a row, and each feature is a unique column.

It is important to clarify the distinction between a **feature** and a **feature model**:

- **Feature (Input Variable)**: An individual numerical value representing an observation, such as the count of a specific word (e.g., the word 'excellent'), the result of a preprocessing rule (e.g., a binary value indicating if a document is fully capitalized), or a score (e.g., a lexicon-based measure of document length).

- **Feature Model (Representation Framework)**: A system, like the bag-of-words (BoW) model, that defines the overall structure of the feature space and dictates which features are created from the raw text (e.g., whether to use single words or N-grams) and how their values are calculated (e.g., raw count or TF-IDF score).

Therefore, **feature engineering** is the deliberate, often creative, process of defining and building the **feature model** to generate the optimal set of individual **features** necessary for the machine learning algorithm to learn successfully.

Table 8.1

Comparison Between Feature, Feature Model, and Feature Engineering

Term	Technical Definition	Example in Text Analysis
Feature	A single, quantifiable, measurable input variable (a column in the matrix).	The frequency count of the word *excellent* (e.g., 5). The presence (1) or absence (0) of the N-gram *not good*. The total length of the document (e.g., 450 words).
Feature Model / Representation	The structured framework or algorithm used to generate the entire matrix of features from unstructured data.	**Bag-of-Words (BoW)**: A model that treats every unique word as a potential feature and records its frequency count. **TF-IDF**: A weighting scheme that modifies the value of a feature (word count) based on its importance across the whole corpus.
Feature Engineering	The systematic process of manually creating and optimizing this feature model to produce the best possible numerical inputs for a classic ML algorithm.	Choosing whether to use **unigrams** or **bigrams** to define the feature set. Deciding whether to apply **stemming** to all words before creating the vector.

8.2.1 The Art of Feature Engineering

Before delving into specific techniques, it is crucial to understand the philosophy of this era. The performance of classic machine learning models was often more dependent on the creativity, domain expertise, and effort invested in crafting the features than on the specific algorithm chosen. This process was, and still is, often described as an 'art.' A practitioner's ability to design features that captured the most salient information for a given task—whether by choosing the right N-gram size, creating custom stop word lists, or incorporating external knowledge sources—was the primary determinant of a project's success (Domingos, 2012). This stands in stark contrast to the modern deep learning paradigm, which is characterized by **representation learning** or **end-to-end learning**. In deep learning, the model is given much rawer input (such as **word embeddings**, **subword encodings**, or **character embeddings**), and it automatically learns the optimal hierarchical features required for the task through its multiple layers. The burden shifts from manual feature design to architectural design and hyperparameter tuning. Understanding this philosophical shift from handcrafted features to learned features is key to understanding the evolution of NLP from the classic statistical era to the modern neural era (LeCun et al., 2015).

8.2.2 Bag-of-Words (BoW)

The simplest and most intuitive **representation** is the **bag-of-words (BoW) model**. This term refers to the earliest and most common method for creating a numerical feature vector of an unstructured text document. It represents a document as an unordered collection of its words, disregarding grammar and word order but keeping track of frequency. For a given **corpus** (a large, systematic collection of texts that serves as the raw, base dataset from which models derive their statistical and linguistic knowledge), a vocabulary of all unique words that come from the corpus is created. Each unstructured text document is then converted into a numerical vector, where the vector length equals the vocabulary size, and the value at each index represents the count of the corresponding word in the document (Harris, 1954; Sahlgren, 2024).

8.2.3 TF-IDF (Term Frequency-Inverse Document Frequency)

A major weakness of BoW is that it treats all words equally. However, common words like 'the' will appear frequently but carry little unique information. **TF-IDF** is a more sophisticated weighting scheme that addresses this by evaluating the importance of a word to a document within a collection or corpus (Jones, 2004).

- **Term Frequency (TF):** Measures how frequently a term appears in a document. It is often normalized to prevent a bias towards longer documents.

- **Inverse Document Frequency (IDF):** Measures how much information a word provides. It is calculated as the logarithm of the number of documents in the corpus divided by the number of documents that contain the word. Rare words have a high IDF score; common words have a low one.

The final TF-IDF score for a word in a document is the product of its TF and IDF scores. This gives more weight to words that are frequent in a specific document but rare across the entire corpus, making them better signals for classification (Salton & Buckley, 1988; Peng et al., 2014).

8.2.4 N-gram Features

The BoW and TF-IDF models' biggest weakness is that they ignore word order. To overcome this and capture some local context, the **N-gram** feature set is incorporated into the vector space model. Instead of a vocabulary of single words (unigrams), the feature model is configured to use contiguous word sequences, or N-grams, as the basic unit. For sentiment analysis, a model using N-gram features can learn that the specific sequence "not good" (a bigram) is a strong indicator of negative sentiment, a nuance that a purely unigram model would miss as it treats 'not' and 'good' as separate, independent features (Pang et al., 2002).

8.3 Classic Machine Learning Algorithms for NLP

With raw text successfully converted into a numerical feature matrix via feature engineering techniques such as bag-of-words and TF-IDF (Section 8.2), the next critical phase is to apply a machine learning algorithm to perform the desired task. Classic machine learning was originally applied to NLP primarily to solve the major problem of **text classification**; thus, this section focuses on the three dominant types of supervised learning classification models that excelled during the classical era: probabilistic, linear, and maximum-margin classifiers. These algorithms enabled the field to solve crucial document-level problems, including content moderation (e.g., spam detection), sentiment analysis, and topic classification/document routing. By mastering these techniques, practitioners established the industry benchmarks that later neural models would seek to surpass.

8.3.1 Probabilistic/Generative Models: Naive Bayes

The Naive Bayes classifier is a simple yet powerful probabilistic model grounded in Bayes' theorem. It has become a standard baseline for text classification due to its efficiency and strong performance, particularly in early spam-filtering research (Sahami et al., 1998; Sun, 2025). For text classification, this model calculates the probability that a document belongs to a specific category, such as spam or non-spam, based on the words it contains. It operates under a 'naive' assumption: that all features (or words) are conditionally independent of one another, given the class. For instance, the Naive Bayes model assumes that if we know a document is classified as 'spam,' the likelihood of finding the word 'click' is independent of the likelihood of finding the word 'here' (a key theoretical flaw of the model). While this assumption is rarely valid in language, as the word 'learning' often heavily relies on being near 'machine,' the Naive Bayes classifier still performs remarkably well in practice (McCallum & Nigam, 1998; Xu, 2018).

How it Works: From Bayes' Theorem to Log Probabilities

To understand how Naive Bayes works, we start with Bayes' theorem, which allows us to calculate the probability of a class given a document:

Equation 8.1

Bayes' Theorem

$$P(\text{class}|\text{document}) = \frac{P(\text{document}|\text{class}) \times P(\text{class})}{P(\text{document})}$$

Our goal is to find the class that maximizes this probability. Since P(document) is the same for all classes, we can ignore it and focus on maximizing the numerator. The term P(class) is the prior probability of the class, which is easily calculated from the training data (e.g., the percentage of spam emails). The challenge is calculating P(document | class). This is where the 'naive' assumption of conditional independence comes in. We assume that the probability of the document, given the class, is simply the product of the probabilities of each individual word in the document, given that class (Lewis, 1998).

This simplification is what makes the model computationally tractable. However, a critical implementation detail arises: multiplying many small probabilities (since the probability of any single word is low) can lead to a floating-point underflow error, where the result becomes too small for the computer to represent and rounds to zero. To solve this, practitioners work in **log space**. Instead of multiplying probabilities, we sum their logarithms. The decision rule becomes finding the class that maximizes the sum of the log probabilities, which is numerically stable and produces the same result (Jurafsky & Martin, 2023).

8.3.2 Discriminative Models: Logistic Regression

Logistic regression is another powerful and highly interpretable linear model used for classification. Unlike Naive Bayes, it does not make strong independence assumptions about the features. Instead, it learns a weight for each feature (e.g., each word in a TF-IDF vector) that represents how much that feature contributes to a positive or negative classification (Hastie et al., 2009). A positive weight might indicate an association with a positive class (e.g., 'excellent'), while a negative weight would indicate an association with a negative class (e.g., 'awful'). To make a prediction, the model calculates a weighted sum of the document's features and passes it through a sigmoid (or 'logistic') function, which compresses the output to a probability between 0 and 1.

The Geometric View and Regularization

Figure 8.1

The Linear Classifier: Finding a Hyperplane to Separate Data

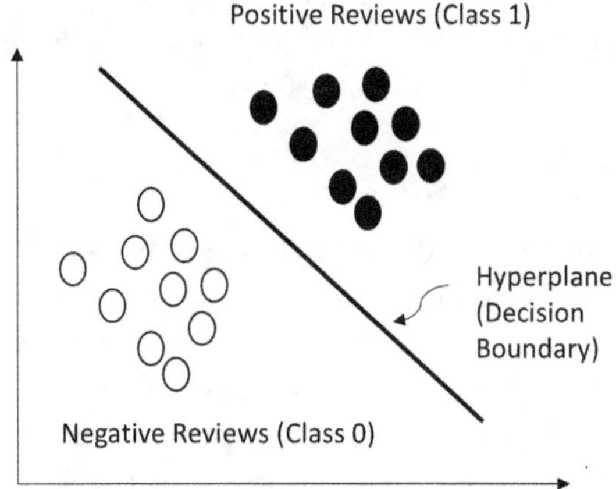

While Naive Bayes is a probabilistic model, logistic regression is best understood geometrically. It learns a linear decision boundary, or **hyperplane**, that separates the classes in the high-dimensional feature space. For a given document vector, the model determines which side of the hyperplane it falls on to classify it. This provides a different and often more powerful way of modeling the problem (Ng & Jordan, 2002).

However, when working with very high-dimensional data, such as text (where a TF-

IDF vector can have tens of thousands of dimensions), there is a significant risk of **overfitting**. The model can learn to perfectly separate the training data but fail to generalize to new, unseen data. To combat this, a technique called **regularization** is used. Regularization adds a penalty term to the model's loss function that discourages it from learning overly complex models with large weights. The two most common types are **L2 regularization (Ridge)**, which penalizes the sum of the squared weights and encourages smaller, more diffuse weights, and **L1 regularization (Lasso)**, which penalizes the sum of the absolute values of the weights. A key property of L1 regularization is that it can force the weights of less important features to become exactly zero, effectively performing a type of automatic feature selection, which can be very useful for interpretability (Tibshirani, 1996).

8.3.3 Discriminative Models: Support Vector Machines (SVMs)

Support vector machines (SVMs) are a class of powerful supervised learning models that became extremely popular for text classification tasks due to their high accuracy and theoretical elegance (Joachims, 1998). Like logistic regression, an SVM is a linear classifier that learns a hyperplane to separate two classes. However, its approach to finding the optimal hyperplane is different and is based on the principle of maximizing the margin (Cortes & Vapnik, 1995).

The Maximum Margin Classifier

The core intuition behind the SVM is that, among all possible hyperplanes that can separate the two classes, the best one is the one that is farthest from the closest data points of either class. This distance between the hyperplane and the nearest data point is called the **margin**. The SVM's objective is to find the hyperplane that creates the **maximum margin**. This creates the widest possible 'street' between the two classes, making the classifier more robust to new data. The data points on the edges of this margin are called **support vectors**. A key property of SVMs is that the decision boundary is determined *only* by these support vectors; all other data points are irrelevant. This makes SVMs particularly efficient and effective in high-dimensional spaces, as the decision is based on a small subset of the training data (Burges, 1998).

The Kernel Trick: Achieving Non-Linearity

The power of SVMs is not limited to linearly separable data. Many real-world datasets cannot be separated by a simple straight line. This is where the **kernel trick** comes in. The kernel trick is a mathematically elegant and computationally efficient method for creating a non-linear classifier (Boser et al., 1992). The core idea is to project the data into a higher-dimensional space where it *becomes* linearly separable. For example, imagine a 1D dataset that is not linearly separable. By applying a simple mapping function,

Equation 8.2

Mapping Function

$$x \rightarrow (x, x^2)$$

we can project the data into a 2D space where a simple line can now separate the classes. The 'trick' of a kernel function (like the polynomial or radial basis function kernel) is that it allows the SVM to calculate the dot products between vectors in this higher-dimensional space *without ever having to explicitly compute the new coordinates*. This provides the power of a non-linear classifier with the computational efficiency of a linear one, and it was a key reason for the SVM's dominance in text classification for many years (Hofmann et al., 2008).

Figure 8.2

The SVM Kernel Trick: Making Non-Linear Data Separable

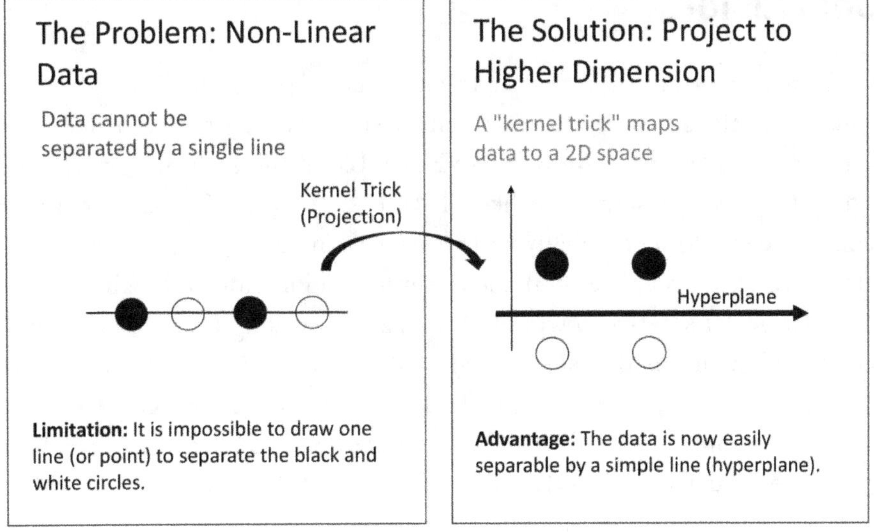

8.4 Limitations and the Motivation for Deep Learning

While powerful, these classic models share a set of fundamental limitations that created a clear and compelling need for a new approach (Bengio et al., 2003):

1. **Dependence on Manual Feature Engineering:** The performance of these models is entirely dependent on the quality of the handcrafted features (BoW, TF-IDF, N-grams). This process is time-consuming, requires significant domain expertise, and may not capture the most optimal representations (Aggarwal & Zhai, 2012).

2. **Inability to Capture Semantics:** Feature representations such as BoW and TF-IDF are based on word co-occurrence rather than meaning. They treat words as discrete, independent symbols. In this view, the words 'car,' 'automobile,' and 'vehicle' are as different from each other as they are from 'apple.' The models have no inherent understanding that these words are semantically related (Turney & Pantel, 2010).

3. **The Problem of Sparsity:** Text vocabularies can be enormous, leading to extremely high-dimensional feature vectors. Since any given document only contains a small fraction of the total vocabulary, these vectors mostly contain zeros. This sparsity can make it difficult for models to learn effectively.

8.5 Conclusion

The classic machine learning era was a transformative period for NLP, demonstrating that robust, scalable language-understanding systems could be built by learning statistical patterns from data. Models like Naive Bayes and support vector machines, powered by feature representations like TF-IDF, set high-performance benchmarks and remain valuable tools for many text classification tasks today (Aggarwal & Zhai, 2012). However, their reliance on manual feature engineering and inability to grasp semantic relationships between words created a clear ceiling on their capabilities. It is precisely these limitations that set the stage for the deep learning revolution. The following chapters will explore how neural networks, with their ability to learn rich, dense representations directly from data, provided the answer to these challenges, ushering in the modern era of NLP (LeCun et al., 2015).

8.6 References

Aggarwal, C. C., & Zhai, C. (2012). A survey of text classification algorithms. In C. C. Aggarwal & C. Zhai (Eds.), *Mining text data* (pp. 163-222). Springer. https://doi.org/10.1007/978-1-4614-3223-4_6

Bengio, Y., Ducharme, R., Vincent, P., & Jauvin, C. (2003). A neural probabilistic language model. *Journal of Machine Learning Research, 3*, 1137–1155.

Boser, B. E., Guyon, I. M., & Vapnik, V. N. (1992). A training algorithm for optimal margin classifiers. *Proceedings of the Fifth Annual Workshop on Computational Learning Theory*, 144–152. https://doi.org/10.1145/130385.130401

Burges, C. J. (1998). A tutorial on support vector machines for pattern recognition. *Data Mining and Knowledge Discovery, 2*(2), 121-167. https://doi.org/10.1023/A:1009715923555

Cortes, C., & Vapnik, V. (1995). Support-vector networks. *Machine Learning, 20*(3), 273–297. https://doi.org/10.1007/BF00994018

Domingos, P. (2012). A few useful things to know about machine learning. *Communications of the ACM, 55*(10), 78–87. https://doi.org/10.1145/2347736.2347755

Harris, Z. S. (1954). Distributional structure. *Word, 10*(2-3), 146–162. https://doi.org/10.1080/00437956.1954.11659520

Hastie, T., Tibshirani, R., & Friedman, J. (2009). *The elements of statistical learning: Data mining, inference, and prediction* (2nd ed.). Springer. https://doi.org/10.1007/978-0-387-84858-7

Hofmann, T., Schölkopf, B., & Smola, A. J. (2008). Kernel methods in machine learning. *The Annals of Statistics, 36*(3), 1171–1220. https://doi.org/10.1214/009053607000000677

Joachims, T. (1998). Text categorization with support vector machines: Learning with many relevant features. *European Conference on Machine Learning*, 137-142. https://doi.org/10.1007/BFb0026683

Jones, K. S. (2004). A statistical interpretation of term specificity and its application in retrieval. *Journal of Documentation, 60*(5), 493–502. https://doi.org/10.1108/00220410410560573

Jurafsky, D., & Martin, J. H. (2023). *Speech and language processing* (3rd ed.). Prentice Hall.

LeCun, Y., Bengio, Y., & Hinton, G. (2015). *Deep learning. Nature, 521*(7553), 436–444. https://doi.org/10.1038/nature14539

Lewis, D. D. (1998). Naive (Bayes) at forty: The independence assumption in information retrieval. *European conference on machine learning*, 4–15. https://doi.org/10.1007/BFb0026666

McCallum, A., & Nigam, K. (1998). A comparison of event models for Naive Bayes text classification. *AAAI-98 Workshop on Learning for Text Categorization*, 41–48.

Ng, A. Y., & Jordan, M. I. (2002). On discriminative vs. generative classifiers: A comparison of logistic regression and Naive Bayes. *Advances in Neural Information Processing Systems*, 14.

Pang, B., Lee, L., & Vaithyanathan, S. (2002). Thumbs up? Sentiment classification using machine learning techniques. *Proceedings of the ACL-02 Conference on Empirical Methods in Natural Language Processing*, Vol. 10, 79–86. https://doi.org/10.3115/1118693.1118704

Peng, T., Liu, L., & Zuo, W. (2014). PU text classification enhanced by term frequency-inverse document frequency-improved weighting. *Concurrency and Computation, 26*(3), 728–741.

https://doi.org/10.1002/cpe.3040

Sahami, M., Dumais, S., Heckerman, D., & Horvitz, E. (1998). A Bayesian approach to filtering junk e-mail. AAAI-98 workshop on learning for text categorization, 55–62.

Sahlgren, M. (2024). Distributional legacy: The unreasonable effectiveness of Harris's distributional program. *Word (Worcester)*, *70*(4), 246–257. https://doi.org/10.1080/00437956.2024.2414515

Salton, G., & Buckley, C. (1988). Term-weighting approaches in automatic text retrieval. *Information Processing & Management*, *24*(5), 513–523. https://doi.org/10.1016/0306-4573(88)90021-0

Salton, G., Wong, A., & Yang, C. S. (1975). A vector space model for automatic indexing. *Communications of the ACM*, *18*(11), 613–620. https://doi.org/10.1145/361219.361220

Sun, Z. (2025). The effect of using Naive Bayes to detect spam email. *ITM Web of Conferences*, *70*, 3027. https://doi.org/10.1051/itmconf/20257003027

Tibshirani, R. (1996). Regression shrinkage and selection via the lasso. *Journal of the Royal Statistical Society: Series B (Methodological)*, *58*(1), 267–288. https://doi.org/10.1111/j.2517-6161.1996.tb02080.x

Turney, P. D., & Pantel, P. (2010). From frequency to meaning: Vector space models of semantics. *Journal of Artificial Intelligence Research*, *37*, 141–188. https://doi.org/10.1613/jair.2934

Wong, S. K. M., & Ziarko, W. (1985). On generalized vector space model in information retrieval. *Fundamenta Informaticae*, *8*(2), 253–267. https://doi.org/10.3233/FI-1985-8207

Xu, S. (2018). Bayesian Naïve Bayes classifiers to text classification. *Journal of Information Science*, *44*(1), 48–59. https://doi.org/10.1177/0165551516677946

8.7 Glossary

Bag-of-Words (BoW): A simple text representation model that describes a document by the frequency of its words, disregarding grammar and word order.

Bayes' Theorem: A mathematical formula for determining conditional probability. It is the foundation of the Naive Bayes classifier.

Compositionality: The principle that the meaning of a complex expression (like a sentence) is determined by the meanings of its constituent parts and the rules used to combine them.

Corpus: A large and structured collection of texts, used for training and evaluating NLP models.

Feature (Input Variable): An individual numerical value representing an observation in the feature matrix, such as the count of a specific word or the result of a preprocessing rule.

Feature Engineering: The process of using domain knowledge to create features that make machine learning algorithms work. In classic NLP, this involves converting text into numerical vectors.

Feature Model (Representation Framework): A system, like the bag-of-words (BoW) model, that defines the overall structure of the feature space and dictates which features are created from the raw text and how their values are calculated.

Hyperplane: A subspace whose dimension is one less than that of its ambient space. In SVMs, it's the decision boundary that separates classes.

Kernel Trick: A technique used in SVMs that allows the model to learn complex, non-linear decision boundaries by projecting the data into a higher-dimensional space where it becomes linearly separable.

Logistic Regression: A Discriminative statistical algorithm used for binary classification that learns a linear decision boundary between two classes.

Margin: In an SVM, the distance between the decision hyperplane and the nearest data points (the support vectors) from either class.

N-gram: A contiguous sequence of n items (typically words) from a given sample of text. Unigrams (n=1), bigrams (n=2), and trigrams (n=3) are common types.

Naive Bayes: A Generative probabilistic classifier based on Bayes' theorem with a 'naive' assumption that all features are independent of one another.

Overfitting: A problem in machine learning where a model learns the training data too well, including its noise and random fluctuations, and thus fails to generalize to new, unseen data.

Regularization: A technique used to prevent model overfitting by adding a penalty term to the optimization objective.

Support Vector Machine (SVM): A discriminative supervised learning model that finds an optimal hyperplane that maximizes the margin between two classes.

Support Vectors: The data points that lie closest to the decision boundary in an SVM and are the most difficult to classify. They are the critical points that define the margin.

TF-IDF (Term Frequency-Inverse Document Frequency): A numerical statistic that reflects how important a word is to a document in a collection or corpus. It combines term frequency (how often a word appears in a document) with inverse document frequency (how rare the word is across all documents).

Vector Space Model: A model that represents text documents as vectors in a multi-dimensional space.

8.8 Review Questions and Discussion Topics

1. **Feature Representation:** Given the sentence "The future of AI is the future of humanity," create the bag-of-words vector representation. What information is lost in this representation? How would using bigram features in addition to unigram (single-word) features help?

2. **Naive Bayes Assumption:** The core assumption of the Naive Bayes model is that features are conditionally independent. Explain why this assumption is 'naive' for natural language. Provide a specific example of two words that are clearly not independent.

3. **Model Selection:** You are tasked with building a system to categorize customer support emails into categories such as "billing inquiry," "technical support," and "general feedback." You have a small dataset and limited computational resources. Which classic ML model from this chapter would you choose as a starting point, and why? Justify your choice based on the model's characteristics.

4. **The Semantic Gap:** Explain the "semantic gap" that classic models like SVMs face due to their reliance on features like TF-IDF. How does their failure to model compositionality (as defined in the glossary) contribute to this gap, and

how did this limitation motivate the development of deep learning?

5. **SVMs and the Kernel Trick:** What is the primary motivation for using the kernel trick in an SVM? Why is it more computationally efficient than manually transforming the features into a higher-dimensional space?

6. **Overfitting and Regularization:** Section 8.3.2 discusses regularization as a technique to prevent overfitting. Explain what overfitting means in the context of a text classification model. Why is a technique like L1 or L2 regularization particularly important when working with high-dimensional text data represented by TF-IDF vectors?

Chapter 9

Machine Translation: The Quest for Meaning via Intermediate Representation

9.1 Introduction

Machine translation (MT) stands as one of the oldest and most ambitious goals of computer science: the automated translation of text or speech from a source language to a target language. The history of MT is a long one, marked by early optimism followed by significant challenges that have driven much of the innovation in NLP. Modern online systems now provide access to documents in over 100 languages for hundreds of millions of users, processing hundreds of billions of words daily. While not yet flawless, the quality is often sufficient for comprehension, and for certain language pairs with vast training data (e.g., Spanish-English), performance in specialized domains can approach that of a human translator.

At the heart of this endeavor lies a fundamental challenge: languages are not simply word-for-word ciphers of one another. They have different grammars, word orders, and cultural contexts. A successful translation must therefore preserve the *meaning* of the source text, not just its words. For decades, the central question in MT research has been how to represent this meaning in a way that a machine can manipulate. A dominant strategy has been the use of an **intermediate representation**—an abstract, often language-neutral structure that captures the 'meaning' of the source text. The goal is to analyze the source sentence, convert it to this intermediate form, and then generate the target sentence from that representation, a concept often referred to as the 'interlingua' approach. This strategy attempts to separate the problem of understanding from that of generation, allowing the system to reason at a deeper level before beginning generation in the target language.

This chapter explores the evolution of machine translation, tracing the path from explicit, human-engineered structures to the statistical models that dominated for decades, and finally to the implicit, learned representations that power today's most advanced systems. We will see how each approach attempts to solve the core problem of translation: how to preserve meaning when the form must change

9.2 The Vauquois Triangle: A Conceptual Map of MT

To understand the different philosophies of machine translation, it is helpful to visualize them using the **Vauquois triangle**, a classic conceptual diagram in the field (Vauquois, 1968). The triangle illustrates the trade-off between the depth of linguistic analysis required and the complexity of the translation process. The bottom of the triangle represents the source and target languages, and the path between them represents the translation process.

The different levels of the triangle are:

- **Direct Translation (Base of the Triangle):** This is the shallowest approach, moving directly from the source text to the target text. It is essentially a word-for-word replacement using a bilingual dictionary, with some simple morphological and reordering rules. This approach is fast but brittle, often producing grammatically incorrect and nonsensical output.

- **Syntactic Transfer (Moving Up):** This approach involves performing a full syntactic parse of the source sentence (e.g., into a constituency or dependency tree). It then applies a set of transfer rules to convert the source-language parse tree into a target-language parse tree. Finally, it generates the target sentence from this new tree. This is more robust than direct translation but requires complex, hand-engineered transfer rules for each language pair.

- **Semantic Transfer:** This level involves an even deeper analysis, aiming to produce a more abstract, language-independent semantic representation (like the semantic roles discussed in Chapter 3). The transfer rules at this level are simpler and more general, as they operate on meaning rather than syntax.

- **Interlingua (Peak of the Triangle):** This is the most ambitious approach. It involves analyzing the source text into a completely language-neutral 'interlingua'—a formal representation of pure meaning. From this interlingua, the target text can be generated into any language, provided a generator exists for that language. This eliminates the need for language-pair-specific transfer rules but requires solving the immense challenge of creating a truly universal representation of meaning

Figure 9.1

The Vauquois Triangle: Conceptual Approaches to Machine Translation

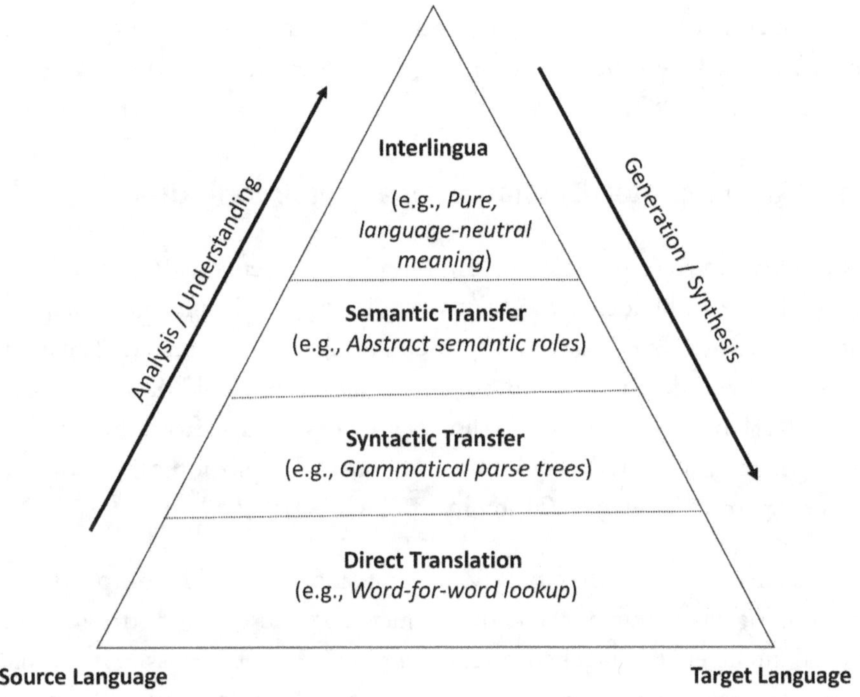

Note: A conceptual map of machine translation approaches. This diagram visually organizes the different theories of machine translation to show how they relate to one another in terms of complexity and depth of analysis.

The Vauquois triangle elegantly shows that the more analytical work you do on the source language (climbing up the left side), the less complex the transfer step becomes, until at the peak, transfer is eliminated entirely. The history of MT can be seen as a continuous effort to climb higher up this triangle (Hutchins & Somers, 1992).

9.3 The Symbolic Era: Rule-Based Representations

Early symbolic MT systems relied on creating explicit, human-designed representations of a sentence's grammatical structure, a process heavily influenced by linguistic theory. These approaches correspond to the 'transfer' levels of the Vauquois triangle. They involved deep syntactic and semantic analysis, using techniques like constituency parsing, dependency parsing, and semantic role labeling to create an intermediate representation that could then be converted into the target language (Nirenburg, 1987).

9.3.1 The "AI-Complete" Challenge of a True Interlingua

The peak of the Vauquois triangle, the interlingua, represents a long-standing and profound goal in AI. The idea is to create a single, universal, language-independent representation of meaning. If such a representation could be created, translation would be 'solved'; to add a new language to a system, one would only need to build a parser from that language into the interlingua and a generator from the interlingua into that language. This would avoid the combinatorial explosion of creating transfer rules for every pair of languages (Hutchins & Somers, 1992).

However, creating a true interlingua is widely considered an **"AI-complete"** problem, meaning that solving it would be equivalent to creating a strong, human-level artificial intelligence. The challenges are immense. One key issue is **lexical gaps**, where a concept that is lexicalized (represented by a single word) in one language has no direct equivalent in another. The German word *schadenfreude* (the pleasure derived from another's misfortune) or the Danish concept of *hygge* (a sense of cozy contentment) are classic examples. How can an interlingua represent these concepts in a truly universal way? Furthermore, meaning is often deeply embedded in a **cultural context**. A phrase like "due process" has a rich legal and cultural history in common law systems that is difficult to represent in a purely abstract form and may not have an equivalent in other legal traditions. Because of these deep-seated challenges, the pure interlingua approach has never been fully realized, and the field has largely moved towards statistical and neural methods that learn representations from data rather than attempting to engineer them from first principles (Nirenburg et al., 1992).

9.4 The Statistical Era: Phrase-Based SMT

The paradigm that dominated machine translation from the 1990s until the rise of neural networks was **statistical machine translation (SMT)**. SMT abandoned the complex, handcrafted rules of the symbolic era and instead framed translation as a probabilistic problem. The core idea, based on the noisy-channel model from information theory, is to find the most probable target sentence e given a source sentence f (Brown et al., 1993).

The most successful and widely used SMT model was **phrase-based SMT**. Instead of translating word by word, this model works by translating contiguous sequences of words or phrases. The main components of a phrase-based SMT system are (Koehn et al., 2003; Koehn, 2009):

- **A Parallel Corpus:** The system is trained on a massive bilingual corpus comprising millions of sentences professionally translated (e.g., parliamentary proceedings, legal documents).

- **A Translation Model:** This model learns the probability of a source phrase being translated into a target phrase, $P(f|e)$. This is learned by first performing statistical **word alignment** on the parallel corpus to identify which source words correspond to which target words, and then extracting all consistent phrase pairs.

- **A Language Model:** This is a standard N-gram language model (as discussed in Chapter 5) trained solely on the target-language text. Its job is to ensure the fluency of the output by assigning a high probability to grammatically correct, natural-sounding sentences ($P(e)$).

- **A Decoder:** The decoder is a search algorithm. For a given source sentence, there are an astronomical number of possible translations. The decoder's job is to efficiently search through this vast space to find the translation that maximizes the combination of the translation model and language model probabilities. This search process is the most computationally complex part of SMT (Koehn, 2009).

Table 9.1

Main Components of a Phrase-Based SMT System

Component	Role in the SMT Pipeline	Function	Probabilistic Term
Parallel Corpus	Data Source (Input)	Provides millions of professionally translated, aligned sentences to train the probabilistic models.	—
Translation Model	Core Probability	Its function is to learn the statistical likelihood of a source phrase f being correctly translated into a target phrase e. Its mathematical term is $P(f\|e)$	$P(f\|e)$
Language Model	Fluency Constraint	An N-gram model that ensures the output sentence is fluent and grammatically correct in the target language.	$P(e)$
Decoder	Search Algorithm	Its function is to efficiently search the vast space of possible translations to find the one that maximizes the combined probability of the translation model and the language model.	$\arg\max_{e} P(e) \cdot P(f\|e)$

9.5 The Modern Paradigm: Learned Intermediate Representations in NMT

The rise of neural machine translation (NMT), beginning with the sequence-to-sequence framework, and particularly with the advent of the transformer architecture, changed everything (Sutskever et al., 2014; Vaswani et al., 2017). Instead of being explicitly programmed with linguistic rules or statistical tables, NMT models learn their own intermediate representations. The encoder part of a transformer model processes the source text and compresses its meaning into a set of high-dimensional vectors (a numerical representation). The decoder then uses this learned representation to generate the translated text. This vector-based representation is an implicit, high-dimensional, and powerful intermediate representation that has proven far more effective than the explicit structures of the past (Bahdanau et al., 2015).

9.6 Evaluating Machine Translation: The BLEU Score

To measure progress and compare different MT systems, researchers needed an automatic evaluation metric that was fast, inexpensive, and reasonably well correlated with human judgment. The most influential and widely used metric became **BLEU** (bilingual evaluation understudy) (Papineni et al., 2002; Chauhan et al., 2023).

Figure 9.2

How the BLEU Score Works. Measuring Translation Quality with N-gram Precision

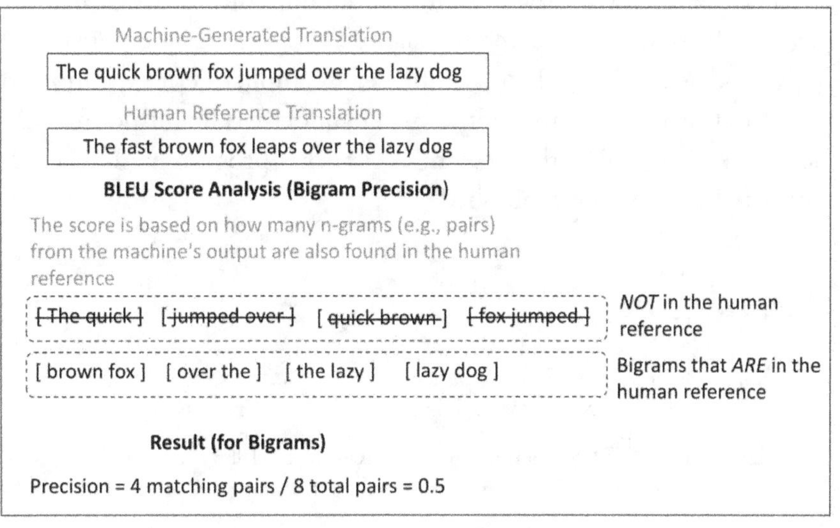

The core idea of BLEU is to measure the precision of N-grams. It compares the N-grams (typically unigrams through 4-grams) in the machine-generated translation to the N-grams in one or more high-quality human reference translations. A translation that shares more N-grams with the human references will receive a higher score. To prevent a model from getting a high score by simply over-generating common words, BLEU uses a 'clipped' precision, meaning that a word is only counted up to the maximum number of times it appears in any single reference translation. It also includes a brevity penalty to penalize translations that are too short. While BLEU has been instrumental in driving much of the field's progress, it has known limitations. It does not account for synonymy (a perfectly good translation using different words will be unfairly penalized) and is insensitive to grammatical correctness if the local N-grams are plausible. For these reasons, while it is an indispensable tool for automatic, day-to-day evaluation during model development, human evaluation remains the gold standard for definitively assessing translation quality (Callison-Burch et al., 2006; Lee et al., 2023).

9.7 Conclusion

The quest for an effective intermediate representation has been a central theme in the history of machine translation. The journey has taken the field from rigid, explicit syntactic structures like constituency parse trees to more flexible dependency grammars, and then to deeper semantic role representations. Each step represented an attempt to create a more abstract, language-neutral representation of meaning. The paradigm shift to neural machine translation, however, has demonstrated that deep learning models can learn their own powerful, implicit representations directly from data, outperforming systems that rely on handcrafted linguistic rules. This ability to learn meaning, rather than having it explicitly programmed, is the defining characteristic of the modern era of NLP.

9.8 References

Bahdanau, D., Cho, K., & Bengio, Y. (2015). *Neural machine translation by jointly learning to align and translate* (arXiv:1409.0473). arXiv. https://doi.org/10.48550/arXiv.1409.0473

Brown, P. F., Della Pietra, S. A., Della Pietra, V. J., & Mercer, R. L. (1993). The mathematics of statistical machine translation: Parameter estimation. *Computational Linguistics, 19*(2), 263–311.

Callison-Burch, C., Osborne, M., & Koehn, P. (2006). Re-evaluating the role of BLEU in machine translation research. *In Proceedings of the 11th Conference of the European Chapter of the Association for Computational Linguistics (EACL 2006)* (pp. 249–256).

Chauhan, S., Daniel, P., Mishra, A., & Kumar, A. (2023). AdaBLEU: A Modified BLEU score for morphologically rich languages. *Journal of the Institution of Electronics and Telecommunication Engineers, 69*(8), 5112–5123. https://doi.org/10.1080/03772063.2021.1962745

Hutchins, W. J. (2007). Machine translation: A concise history. In S. Chan (Ed.), *Computer-aided translation: Theory and practice* (pp. 29–70). Chinese University Press. mt-archive.net

Hutchins, W. J., & Somers, H. L. (1992). *An introduction to machine translation.* Academic Press.

Lee, S., Lee, J., Moon, H., Park, C., Seo, J., Eo, S., Koo, S., & Lim, H. (2023). A Survey on evaluation metrics for machine translation. *Mathematics, 11*(4), 1006. https://doi.org/10.3390/math11041006

Koehn, P. (2009). *Statistical machine translation*. Cambridge University Press.

Koehn, P., Och, F. J., & Marcu, D. (2003). Statistical phrase-based translation. *In Proceedings of the 2003 Conference of the North American Chapter of the Association for Computational Linguistics on Human Language Technology*, Vol. 1 (pp. 48–54).

Nirenburg, S. (Ed.). (1987). *Machine translation: Theoretical and methodological issues*. Cambridge University Press.

Nirenburg, S., Carbonell, J., Tomita, M., & Goodman, K. (1992). *Machine translation: A knowledge-based approach*. Morgan Kaufmann.

Papineni, K., Roukos, S., Ward, T., & Zhu, W. J. (2002). BLEU: A method for automatic evaluation of machine translation. *In Proceedings of the 40th Annual Meeting of the Association for Computational Linguistics* (pp. 311–318). https://doi.org/10.3115/1073083.1073135

Sutskever, I., Vinyals, O., & Le, Q. V. (2014). Sequence to sequence learning with neural networks. *Advances in Neural Information Processing Systems, 27*.

Vaswani, A., Shazeer, N., Parmar, N., Uszkoreit, J., Jones, L., Gomez, A. N., Kaiser, L., & Polosukhin, I. (2017). Attention is all you need. *Advances in Neural Information Processing Systems, 30*.

Vauquois, B. (1968). A survey of formal grammars and algorithms for recognition and transformation in mechanical translation. In *IFIP Congress* (2) (pp. 254–260).

9.9 Glossary

BLEU (Bilingual Evaluation Understudy): An automatic metric for evaluating the quality of machine-translated text by comparing the N-gram precision of the machine's output to high-quality human translations.

Constituency Parse Tree: A hierarchical representation of a sentence's grammatical structure that breaks it down into its component phrases (e.g., noun phrase, verb phrase).

Decoder: In NMT, the part of the model that takes the intermediate representation from the encoder and generates the output text in the target language.

Dependency Structure: A grammatical representation where words are linked based on direct relationships (dependencies), such as a verb and its subject or an adjective and the noun it modifies. It is less rigid about word order than a parse tree.

Encoder: In NMT, the part of the model that reads the input text in the source language and compresses its meaning into a set of high-dimensional vectors.

Interlingua: A theoretical, language-neutral representation of meaning. In the ideal MT system, a source text would be converted to an interlingua and then generated into any target language.

Intermediate Representation: An abstract, language-neutral format used to represent the meaning or structure of a source text. It acts as a bridge between the analysis of the source language and the generation of the target language.

Neural Machine Translation (NMT): An approach to machine translation that uses a single, large artificial neural network to translate text from a source language to a target language.

Parallel Corpus: A collection of texts where each text is available in two or more languages. This is the primary training data for statistical and neural machine translation systems.

Phrase-Based SMT: A type of statistical machine translation that translates by segmenting the source sentence into phrases (sequences of words) and finding the most probable translation for each phrase.

Statistical Machine Translation (SMT): An approach to machine translation that learns to translate by analyzing statistical models whose parameters are derived from the analysis of bilingual text corpora.

Transformer: A specific neural network architecture, introduced in 2017, that has become the dominant model for NMT and many other NLP tasks. It relies heavily on self-attention mechanisms instead of the recurrent structures of earlier models.

9.10 Review Questions and Discussion Topics

1. **Practical Application**: For the English sentence "The fast runner quickly drank the cold water."

   a. Draw a constituency parse tree

   b. Draw a dependency structure diagram

   c. Identify the semantic roles for the predicate 'drank'

2. **Comparative Analysis:** Consider translating from English, a subject-verb-object (SVO) language, to Japanese, a subject-object-verb (SOV) language. These terms describe the typical word order of a sentence (e.g., English: "The student reads the book"; Japanese: "The student the book reads"). Why would a dependency-based representation be theoretically more effective for this task than a constituency parse tree-based representation? Use the sentence "The student reads the book" to illustrate your point.

3 **Ambiguity:** The sentence "I saw the man on the hill with a telescope" is famous for its ambiguity.

   a. Describe at least two different possible meanings for this sentence.

   b. How would the dependency structures differ for each of these meanings?

   c. Discuss why this sentence poses a significant challenge for any type of intermediate representation.

4 **Semantic Roles:** Identify the semantic roles (agent, patient, instrument, location, etc.) in the following sentence: "In the kitchen, the chef carefully sliced the vegetables with a sharp knife."

5 **Implicit vs. Explicit Representation:** The chapter contrasts the explicit, human-designed intermediate representations of the symbolic era with the

implicit, learned representations of the NMT era. Discuss the primary advantages and disadvantages of each approach (consider factors like scalability, interpretability, and debugging).

6   **The Limits of Interlingua:** The idea of a perfect, universal 'interlingua' has been a long-standing goal. Based on the chapter, discuss the primary obstacles to creating a single, explicit intermediate representation that can perfectly capture the meaning of any sentence in any language.

7   **Evaluation:** Section 9.6 introduces the BLEU score as a metric for evaluating machine translation. Discuss the strengths and weaknesses of BLEU. Why is human evaluation still considered the "gold standard" despite the existence of automatic metrics like BLEU?

8   **Paradigm Shifts:** Compare and contrast the three major eras of machine translation discussed in the chapter: rule-based, statistical (SMT), and neural (NMT). What were the primary limitations of each paradigm that motivated the shift to the next?

Chapter 10

Sequence-to-Sequence Models

10.1 Introduction

Sequence-to-sequence (Seq2Seq) learning is a powerful framework for transforming an input sequence into an output sequence, where the lengths of the sequences can be variable. This capability makes it a cornerstone of modern natural language processing (NLP), enabling solutions to complex tasks from machine translation to text summarization.

This chapter will explore the architecture of Seq2Seq models, beginning with their foundational implementation using **recurrent neural networks** (RNNs) and **long short-term memory** (LSTM) networks (Hochreiter & Schmidhuber, 1997). We will examine the core components—the encoder, the decoder, and the crucial context vector that bridges them. We will then delve into the limitations of this initial architecture, such as the "information bottleneck," which led to the development of the revolutionary attention mechanism.

Finally, we will analyze the seminal paper by Sutskever et al. (2014), which set the stage for this new paradigm and provided the first compelling evidence that end-to-end deep learning could tackle complex sequence transformation tasks. Understanding this architecture is essential, as it is the direct predecessor of today's state-of-the-art models and provides the conceptual foundation for more advanced architectures, such as the transformer.

10.2 The Fundamentals of Sequence-to-Sequence Learning

Unlike traditional models that require fixed-size inputs and outputs, Seq2Seq models are designed for the dynamic nature of sequential data like text or speech. The standard architecture is composed of two primary components: an encoder and a decoder (Cho et al., 2014).

- **Encoder:** The encoder's job is to process the entire input sequence, token-by-token. It is typically a **recurrent neural network (RNN), long short-term memory** (LSTM), or **gated recurrent unit** (GRU). At each step, it updates a hidden state, and after processing the final input token, its last hidden state becomes the context vector (Sutskever et al., 2014).
- **Context Vector (or "Thought Vector"):** The context vector is a single, fixed-size vector that aims to encapsulate the meaning of the entire input sequence. It is often referred to as a **"thought vector"**—an analogy that frames the encoder's process as 'reading' and 'thinking' about the source sentence to produce a single, comprehensive thought. This numerical thought must be a dense, information-rich summary (Cho et al., 2014).
- **Decoder:** The decoder is another RNN that takes the context vector as its initial hidden state. Its job is to generate the output sequence one token at a time. The output from each step is fed as the input to the next step, a process known as autoregression, until a special "end-of-sequence" token is produced (Sutskever et al., 2014).

The initial architecture suffered from a significant limitation: the **information bottleneck**. Forcing the encoder to compress the meaning of a long, complex sentence into a single fixed-size vector inevitably led to information loss. This led to the development of the **attention mechanism**, which allows the decoder to "look back" at all of the encoder's hidden states (not just the last one). It learns to dynamically weigh the importance of different parts of the input sequence relevant to the current output word and assign them a higher "attention score," allowing the decoder to focus on different parts of the input sequence as needed. This drastically improves performance, especially for long sequences (Bahdanau et al., 2015).

To make these concepts concrete, the following sections provide a more detailed, step-by-step walkthrough of the information flow in both the standard and attention-based architectures.

10.2.1 Architectural Flow of a Standard Seq2Seq Model

Imagine we are translating the Spanish phrase **"Mi carro es azul"** to **"My car is blue."**

Part 1: The Encoder (Reading and Summarizing)

The encoder's goal is to read the entire input phrase and distill its meaning into a single numerical summary—the context vector.

1. **Reading the First Word**: The encoder receives the first word, '**Mi**.' It processes this word and generates a hidden state, which is a vector of numbers that represents the meaning of 'Mi.'

2. **Reading the Second Word**: The encoder then receives the second word, 'carro.' Crucially, it also receives the hidden state from the previous step. This allows it to understand '**carro**' in the context of 'Mi.' It then produces a new, updated hidden state that now represents the meaning of "Mi carro."

3. **Reading the Third & Fourth Words:** This process repeats sequentially. The encoder processes '**es**' (in the context of "Mi carro"), and then processes 'azul' (in the context of "Mi carro es").

4. **Creating the "Thought Vector"**: After processing the final word ('**azul**'), the encoder's last hidden state is designated as the context vector. This single vector is the encoder's final summary—its 'thought'—that encapsulates the complete meaning of "**Mi carro es azul**."

Part 2: The Decoder (Generating the Translation)

The decoder's goal is to take the "thought vector" and generate the English translation, one word at a time.

1. **Receiving the Summary:** The decoder's initial hidden state is set to be the Context Vector from the encoder. It now 'knows' what the original phrase meant.

2. **Generating the First Word:** The decoder starts with a special <start> token. Using the context vector, it predicts the most likely first word of the English translation: '**My**.'

3. **Generating the Second Word:** The decoder then takes the word it just produced ('My') as its next input. Using its memory and the context, it predicts the next most likely word: '**car**.'

4. **Generating the Third & Fourth Words:** This autoregressive process continues. The decoder takes 'car' as input and predicts 'is'. Then, it takes **'is'** as input and predicts **'blue.'**

5. **Stopping:** Finally, the decoder takes 'blue' as input and predicts a special <end> token, signaling that the translation is complete. The final output is the sequence of generated words: **"My car is blue."**

Figure 10.1

Standard Seq2Seq Model: A clearer, step-by-step visualization of the encoder-decoder flow with a fixed context vector.

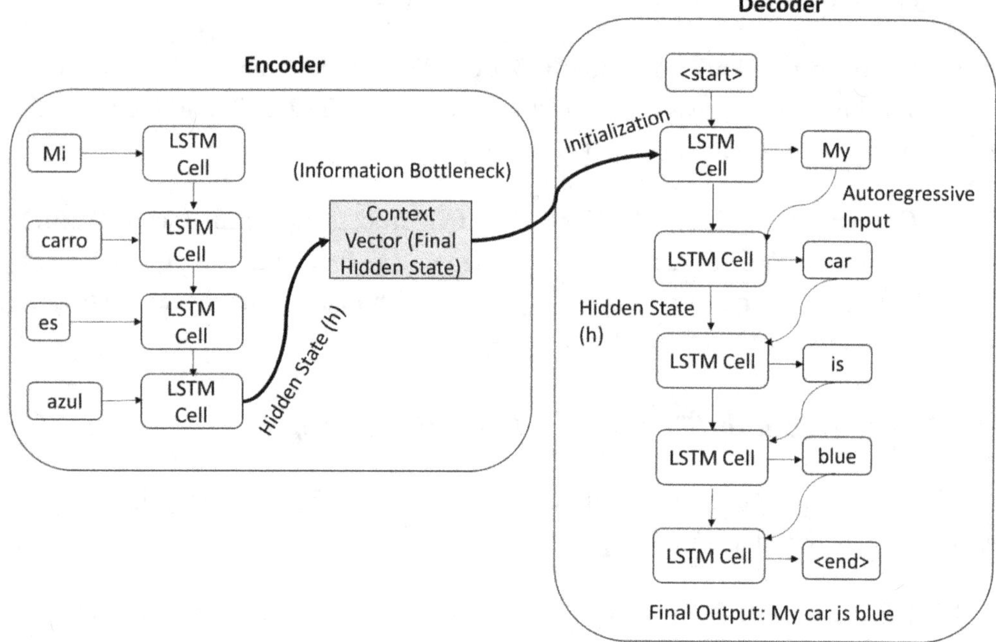

10.2.2 Architectural Flow of a Seq2Seq Model with Attention

The attention mechanism solves the "information bottleneck" by giving the decoder a more powerful way to access the input. Instead of a single final summary, the decoder has access to the encoder's notes at every step.

Part 1: The Encoder (Taking Detailed Notes)

The encoder's job is largely the same, but what it passes to the decoder is different.

1. Reading and Processing: The encoder reads "Mi carro" word by word, producing a hidden state at each step, just like before.

Figure 10.2

Seq2Seq Model with Attention

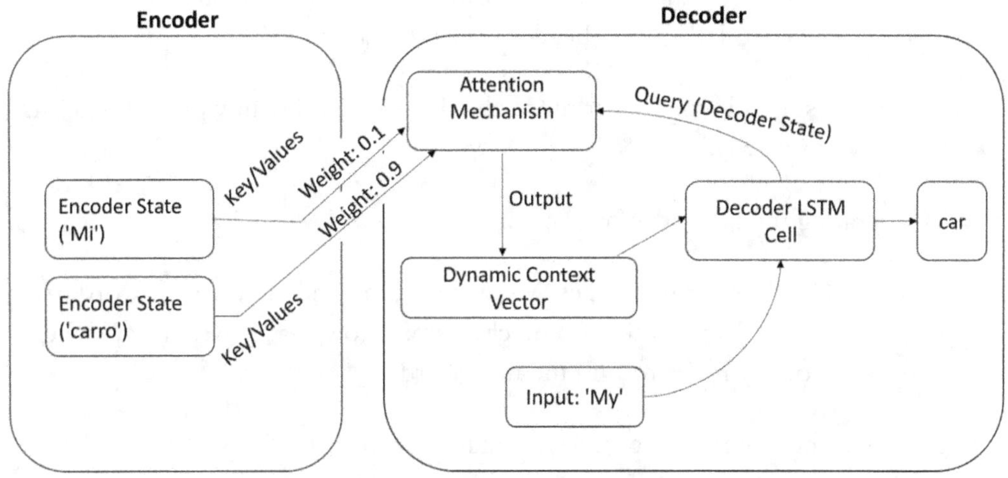

Note: This diagram provides a high-level overview of the attention-based architecture

2. Providing All Hidden States: Instead of discarding the intermediate hidden states, the encoder provides its entire 'notebook' to the decoder. This is a collection of all the hidden states it produced: one representing **'Mi'** in context, and another representing **'carro'** in context.

Part 2: The Decoder (Translating with a Spotlight)

The decoder's process is now more dynamic. At each step, it uses the attention mechanism as a 'spotlight' to focus on the most relevant parts of the input.

1. Generating the First Word ('My'):

   o The decoder is about to generate the first word. It asks the attention mechanism: "To generate the start of the translation, where should I look in the input?"

- o The attention mechanism compares the decoder's current state to all the encoder's hidden states ('Mi' and 'carro'). It calculates that the hidden.

  state for 'Mi' is the most relevant and gives it a high attention weight (e.g., 95%).

- o A new, dynamic context vector is created as a weighted average of the encoder's notes, heavily biased towards 'Mi.'

- o Using this focused context, the decoder confidently predicts the first word: '**My.**'

2. Generating the Second Word ('car'):

- o The decoder now takes 'My' as input to generate the next word. It again asks the attention mechanism: "Now that I've said 'My,' where should I focus to find the next word?"

- o The attention mechanism again compares the decoder's new state with the encoder's notes. This time, it calculates that the hidden state for 'carro' is the most relevant, giving it a high attention weight (e.g., 98%).

- o A new dynamic context vector is created, this time dominated by the meaning of 'carro.'

- o Using this new focused context, the decoder predicts the next word: '**car.**'

This process allows the model to create a direct connection between specific input and output words, making it far more powerful and effective, especially for longer sentences.

10.3 Training and Core Challenges

Seq2Seq models are typically trained end-to-end on a large corpus of paired sequences (e.g., English sentences and their French translations). The model learns by predicting the next correct token in the target sequence, and the error between its prediction and the actual token is used to update the weights of both the encoder and the decoder via backpropagation (Sutskever et al., 2014). However, this process presents several core challenges:

- **Handling Long Sequences:** As discussed, the information bottleneck of the context vector is a major issue for long inputs. The attention mechanism was the primary solution to this problem (Bahdanau et al., 2015).
- **Exposure Bias and Teacher Forcing:** A discrepancy exists between how the model is trained and how it is used. During training, a technique called **teacher forcing** is used to stabilize and improve efficiency. At each step, the decoder is fed the correct "ground truth" token from the target sequence to predict the next one (Williams & Zipser, 1989). This is like learning to ride a bike with training wheels. During inference, however, it must use its own previously generated tokens as input. If the model makes a mistake, that error can propagate and compound, leading to incoherent output sequences. This mismatch between training and inference is known as **exposure bias** (Ranzato et al., 2016).
- **Computational Complexity:** Processing long sequences, especially with recurrent architectures like LSTMs or GRUs that handle tokens sequentially, is computationally intensive in both time and memory. This challenge has been a primary motivator for the development of non-recurrent architectures like the transformer (Vaswani et al., 2017).

Figure 10.3

Training vs. Inference: The Seq2Seq Dilemma

1. Training (with "Teacher Forcing")

The "Answer Key" Method (Fast & Stable)

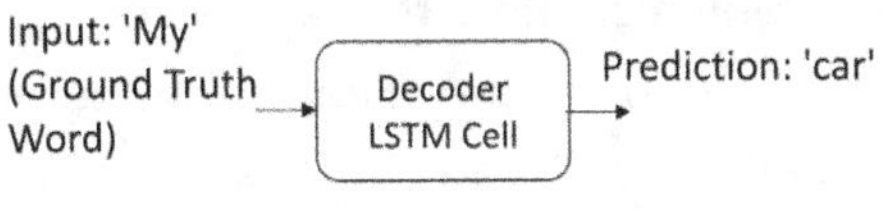

2. Inference (Autoregressive)

The "Real World" Method (Can Drift)

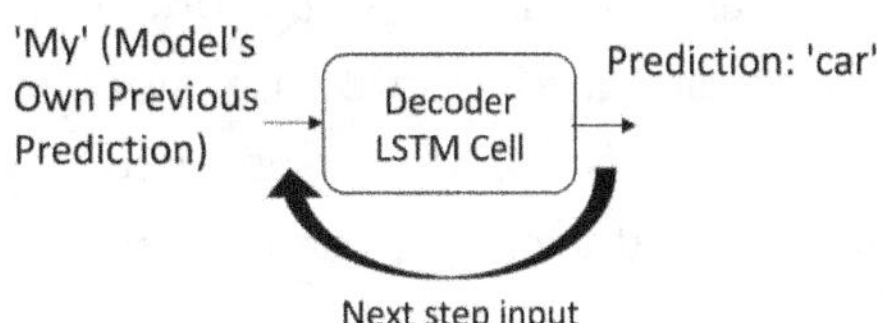

Logic: At every step, the model is fed the correct word from the answer key. This is stable, but the model never learns to recover from its own mistakes.

Logic: The model must use its own last prediction as the input for the next word. An early mistake can cause all future predictions to "drift" off-topic.

10.4 Relevance and Applications in NLP

The Seq2Seq architecture is highly versatile and has been successfully applied to a wide range of NLP tasks that involve transforming one sequence into another. While machine translation was the initial driving application, the framework's flexibility has made it a foundational tool for many other problems (Jurafsky & Martin, 2023):

- **Abstractive Text Summarization:** The input is a long article (a sequence of words), and the output is a shorter, condensed summary (a new sequence of words) (Rush et al., 2015).
- **Conversational AI and Chatbots:** The input is a user's query or statement (a sequence), and the output is the chatbot's response (another sequence) (Vinyals & Le, 2015).
- **Speech Recognition:** The input is a sequence of audio frames, and the output is a sequence of transcribed words (Chorowski et al., 2015).
- **Image Captioning:** A hybrid application where the input is an image (processed by a **convolutional neural network** (CNN) to produce a feature vector, which acts as the initial context vector) and the output is a descriptive caption (a sequence of words) (Vinyals et al., 2015; Baltrušaitis et al, 2019).

10.5 A Foundational Case Study: Sutskever et al. (2014)

To understand how Seq2Seq models moved from theory to practice, we will now analyze the seminal paper *Sequence to Sequence Learning with Neural Networks* by Sutskever et al. (2014). This work provided the first compelling demonstration that a deep LSTM-based model could outperform established statistical machine translation (SMT) systems on a large-scale translation task.

The Problem: DNNs and Variable Sequences. At the time, deep neural networks (DNNs) were already excelling at tasks like image recognition. However, they had a significant limitation: they could only operate on problems where the inputs and outputs were of a fixed, predetermined size. This made them unsuitable for many important NLP tasks, where sentence lengths naturally vary. Sutskever et al. (2014) aimed to solve this exact problem using a novel neural architecture.

The Solution: A Two-LSTM Architecture. The authors proposed a general end-to-end approach to sequence learning that made minimal assumptions about the sequence structure. Their model consisted of two multi-layered LSTMs: one for the input sequence (the encoder) and another for the output sequence (the decoder).

This clean, powerful architecture proved remarkably effective (Sutskever et al., 2014).

Key Innovations and Results The success of the model was not just in the architecture but also in several clever implementation details (Sutskever et al., 2014):

- **Deep LSTMs:** The researchers found that using a deep LSTM with four layers performed significantly better than a shallow one.
- **Reversing the Source Sequence:** They discovered a simple but highly effective trick: reversing the order of the words in the input sentence (e.g., encoding "azul es carro Mi," which is the reverse Spanish sentence for "My car is blue"). This seemingly small change made the optimization problem much easier by creating a shorter 'path' for the error signals during backpropagation between the start of the source and target sentences.
- **Decoding with Beam Search:** Instead of just picking the single most likely word at each step (a 'greedy' approach), they used a **beam search** decoder. This method keeps track of a small number (k, the beam size) of the most probable partial translations at each step. By exploring a wider search space, it is more likely to find a globally optimal sequence, ultimately leading to a more accurate and fluent final translation (Graves, 2012).

Figure 10.4

Seq2Seq Architecture (Sutskever et al., 2014). Visualizing the Deep LSTM Encoder-Decoder with Input Reversal.

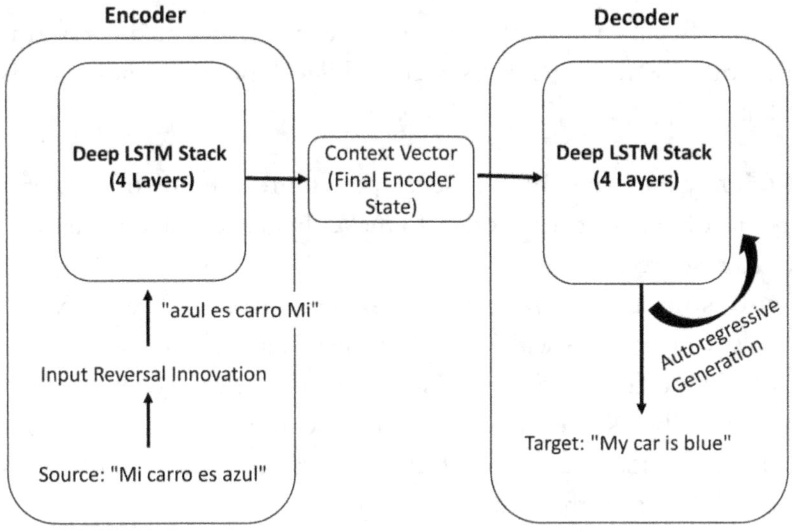

- **Results:** When applied to English-to-French translation, their model, trained on a large dataset, achieved a BLEU score that surpassed the state-of-the-art SMT systems, demonstrating, for the first time, the viability of a purely neural approach to machine translation (Sutskever et al., 2014).

Historical Impact and Legacy The Sutskever et al. (2014) paper was a disruptive innovation that solved the fixed-vector constraint of DNNs for sequential tasks. It demonstrated the power of LSTM-based Seq2Seq learning and set the stage for the next wave of NLP research. However, its reliance on a single context vector, even with the input-reversal trick, was still a bottleneck. This limitation directly motivated the development of the attention mechanism just a few months later, which in turn paved the way for the transformer architectures that define the field today.

10.6 Conclusion

The journey of sequence-to-sequence learning, from its origins in recurrent neural networks to the highly parallelized transformer models of today, represents a paradigm shift in machine learning. We have seen how the simple yet elegant encoder-decoder architecture provides a powerful new approach to problems with variable-length sequences (Sutskever et al., 2014). The limitations of this initial model—namely, the information bottleneck—did not represent a failure but rather a catalyst for further innovation, leading directly to the development of the attention mechanism, a concept that now lies at the heart of modern AI (Bahdanau et al., 2015).

The relevance of Seq2Seq models continues to expand, and many opportunities for future work remain. Key areas for research and development include (Jurafsky & Martin, 2023):

- **Efficiency and Scalability:** Developing more computationally efficient versions of transformer models to make them accessible for tasks with limited resources.
- **Low-Resource Languages:** Creating methods to train effective Seq2Seq models for languages with limited parallel data.
- **Controllable Generation:** Building models that can generate text with specific attributes, such as a particular style, tone, or level of formality.
- **Mitigating Bias:** Developing techniques to reduce the social biases that models learn from their training data.

10.7 References

Bahdanau, D., Cho, K., & Bengio, Y. (2015). *Neural machine translation by jointly learning to align and translate* (arXiv:1409.0473). arXiv. https://doi.org/10.48550/arXiv.1409.0473

Baltrušaitis, T., Ahuja, C., & Morency, L.-P. (2019). Multimodal machine learning: A survey and taxonomy. *IEEE Transactions on Pattern Analysis and Machine Intelligence, 41*(2), 423–443. https://doi.org/10.1109/TPAMI.2018.2798607

Cho, K., Van Merriënboer, B., Gulcehre, C., Bahdanau, D., Bougares, F., Schwenk, H., & Bengio, Y. (2014). Learning phrase representations using RNN encoder-decoder for statistical machine translation. *In Proceedings of the 2014 Conference on Empirical Methods in Natural Language Processing (EMNLP)* (pp. 1724-1734). Association for Computational Linguistics. https://doi.org/10.3115/v1/D14-1179

Chorowski, J. K., Bahdanau, D., Serdyuk, D., Cho, K., & Bengio, Y. (2015). Attention-based models for speech recognition. *Advances in Neural Information Processing Systems, 28.*

Graves, A. (2012). *Sequence transduction with recurrent neural networks.* arXiv preprint arXiv:1211.3711.

Hochreiter, S., & Schmidhuber, J. (1997). Long short-term memory. *Neural Computation, 9*(8), 1735–1780. https://doi.org/10.1162/neco.1997.9.8.1735

Jurafsky, D., & Martin, J. H. (2023). *Speech and language processing* (3rd ed.). Pearson.

Ranzato, M., Chopra, S., Auli, M., & Zaremba, W. (2016). Sequence level training with recurrent neural networks. *In Proceedings of the 4th International Conference on Learning Representations (ICLR).*

Rush, A. M., Chopra, S., & Weston, J. (2015). A neural attention model for abstractive sentence summarization. *In Proceedings of the 2015 Conference on Empirical Methods in Natural Language Processing* (pp. 379-389). Association for Computational Linguistics. https://doi.org/10.18653/v1/D15-1044

Sutskever, I., Vinyals, O., & Le, Q. V. (2014). *Sequence to Sequence Learning with Neural Networks.* https://doi.org/10.48550/arxiv.1409.3215

Vaswani, A., Shazeer, N., Parmar, N., Uszkoreit, J., Jones, L., Gomez, A. N., Kaiser, Ł., & Polosukhin, I. (2017). Attention is all you need. *Advances in Neural Information Processing Systems, 30.*

Vinyals, O., & Le, Q. (2015). *A neural conversational model.* arXiv preprint arXiv:1506.05869.

Vinyals, O., Toshev, A., Bengio, S., & Erhan, D. (2015). Show and tell: A neural image caption generator. *In Proceedings of the IEEE Conference on Computer Vision and Pattern Recognition* (pp. 3156–3164). IEEE. https://doi.org/10.1109/CVPR.2015.7298935

Williams, R. J., & Zipser, D. (1989). A learning algorithm for continually running fully recurrent neural networks. *Neural Computation, 1*(2), 270-280. https://doi.org/10.1162/neco.1989.1.2.270

10.8 Glossary

Attention Mechanism: A technique that allows the decoder to dynamically focus on different parts of the input sequence when generating each token of the output, overcoming the bottleneck of a single context vector.

Autoregression: A model property where the prediction for the current time step is generated based on the outputs from previous time steps. The model's own output is fed back in as input for the next step.

Beam Search: A decoding algorithm used during inference that explores multiple probable output sequences simultaneously, rather than just choosing the single best word at each step.

BLEU (Bilingual Evaluation Understudy): An automatic metric for evaluating the quality of machine-translated text by comparing the N-gram precision of the machine's output to high-quality human translations.

Context Vector: A numerical representation (a vector) that serves as a summary of the encoder's input. It acts as the initial state for the decoder.

Convolutional Neural Network (CNN): A deep learning architecture designed for processing structured data, like images, audio, and video. Its main innovation is convolutional layers that learn hierarchical features from raw input. In natural language processing (NLP) tasks such as image captioning, the CNN acts as a visual encoder, extracting key visual features into a fixed-size vector that guides the text-generating decoder.

Decoder: The component of a Seq2Seq model that takes the context vector and generates the output sequence one token at a time.

Encoder: The component of a Seq2Seq model that reads the input sequence and compresses its information into a fixed-size context vector.

Exposure Bias: The discrepancy between training (where the decoder sees the correct "ground truth" inputs) and inference (where the decoder must use its own, potentially flawed, outputs as input).

Gated Recurrent Unit (GRU): A simplified version of the LSTM with fewer parameters, which often performs comparably.

Information Bottleneck: A limitation in the standard Seq2Seq architecture where the entire meaning of a long input sequence must be compressed into a single, fixed-size context vector, often resulting in information loss.

Long Short-Term Memory (LSTM): An advanced type of RNN architecture that uses 'gates' to control the flow of information, allowing it to learn long-range dependencies and avoid the vanishing gradient problem.

Recurrent Neural Network (RNN): A type of neural network designed to work with sequential data, where connections between nodes form a directed graph along a temporal sequence.

Sequence-to-Sequence (Seq2Seq) Model: An architecture that uses an encoder to process an input sequence and a decoder to generate an output sequence, allowing for variable-length inputs and outputs.

Teacher Forcing: A training technique for sequence generation models where the decoder receives the ground-truth output from the previous time step as input, rather than its own prediction.

10.9 Review Questions and Discussion Topics

1. **The Information Bottleneck:** Explain in your own words why the fixed-size context vector in the original Seq2Seq model is referred to as an "information bottleneck." How does the attention mechanism specifically solve this problem?

2. **Input Reversal:** In the Sutskever et al. (2014) paper, the researchers found that reversing the order of the source sentence significantly improved performance. Why did this simple trick have such a powerful effect? What does this reveal about how information flows during the training of an RNN?

3. **Architectural Trade-offs:** While this chapter focused on LSTM-based models, the state of the art is now the transformer. Based on the challenges discussed (e.g., computational complexity), what advantages might a transformer's non-recurrent, parallel processing approach have over a sequential LSTM-based model? What might be some potential disadvantages?

4. **Brainstorming Applications:** Beyond the examples given in the chapter (translation, summarization, dialogue, speech recognition, image captioning), brainstorm two new and creative applications that could be framed as a sequence-to-sequence problem. For each, define what the input and output sequences would be.

5. **Decoding Strategies:** The case study on Sutskever et al. (2014) highlights the use of beam search for decoding. Contrast this approach with a simpler 'greedy'

decoding strategy. What are the advantages of using beam search, and what is the primary trade-off involved?

6. **Training vs. Inference:** Section 10.3 discusses the concept of "exposure bias." Explain the difference between training a Seq2Seq model with Teacher Forcing and using it for inference. Why does this discrepancy lead to exposure bias, and how might it affect the quality of the generated sequence?

Chapter 11

Neural Machine Translation

11.1 Introduction: The Challenge of Context in Translation

Consider a simple English sentence: "The bill was passed." A machine translation system might render this in Spanish as "La factura fue aprobada" (The invoice was approved) or "El proyecto de ley fue aprobado" (The legislative bill was approved). Without a broader context, both translations are plausible. This ambiguity highlights a fundamental challenge in machine translation (MT): the critical importance of context. For decades, translation systems struggled with this, often producing literal but nonsensical translations, especially for long or complex sentences. How can we design a system that not only translates words but also 'understands' the surrounding context to make more intelligent choices?

This chapter explores the evolution of **neural machine translation (NMT)**, a paradigm that revolutionized the field by leveraging deep learning to address this problem. We begin by examining the initial **RNN encoder-decoder framework** and its primary limitation—the **information bottleneck**, as introduced conceptually in Chapter 10. We then present the breakthrough solution proposed by Bahdanau et al. (2015): a **bidirectional RNN (BiRNN) encoder-decoder** equipped with a dynamic **attention mechanism**. This innovation allowed the model to dynamically focus on the most relevant parts of a source sentence during translation, dramatically improving fluency and accuracy, and setting a new standard for machine translation quality.

11.2 The RNN Encoder-Decoder: A Fixed-Context Approach

As we discussed in the previous chapter, the foundational architecture for NMT is the **RNN encoder-decoder framework**. This model provided the first end-to-end neural solution for sequence-to-sequence tasks like machine translation. The architecture consists of two primary components: an **encoder** that processes the input sentence and a **decoder** that generates the translated output sentence.

The **encoder**, an RNN, processes the input sequence word by word. At each time step t, it updates its hidden state, h_t, based on the current input word, x_t, and its previous hidden state, h_{t-1}. While effective for short sequences, standard RNNs suffer from the **vanishing gradient problem**. While standard recurrent neural networks (RNNs) can effectively handle short sequences, they struggle with the **vanishing gradient problem** in longer sequences. During training, error signals—represented by the **gradients of the loss function**—are used to continuously adjust the model's internal weights to maintain accurate outputs. However, as these error signals are propagated backward through the network over many time steps in a long sentence, they can become exponentially small. This diminishing effect makes it difficult for the model to learn relationships between distant words (Bengio et al., 1994).

Technical Deep Dive: The RNN Hidden State

The relationship that defines the hidden state update is defined by the function:

Equation 11.1

The RNN Hidden State Update

$$h_t = f(W_{hh}h_{t-1} + W_{xh}x_t)$$

- h_t: The new hidden state at the current time step t. This is the output of the current RNN cell and will be used as input for the next time step. It's a vector.
- h_{t-1}: The previous hidden state from time step t-1. This is the 'memory' from the past that the network uses.
- x_t: The input at the current time step t. This is the new piece of

data being fed into the network (e.g., a word embedding in a sentence). It's a vector.

- W_{hh}: The weight matrix for the recurrent (hidden-to-hidden) connection. It determines the extent to which the previous hidden state (h_{t-1}) influences the new one. This matrix learns during training.
- W_{xh}: The weight matrix for the input (input-to-hidden) connection. It determines the extent to which the current input (x_t) affects the new hidden state. This matrix learns during training.
- $f(...)$: A non-linear activation function, such as the hyperbolic tangent (tanh) or rectified linear Unit (ReLU). This function is applied element-wise to the result of the matrix operations. It allows the network to learn more complex patterns than a simple linear combination would.

For this task, a more sophisticated variant of the RNN, the **long short-term memory (LSTM)** network, is often employed. LSTMs are explicitly designed to combat the vanishing gradient problem inherent in standard RNNs, using a series of 'gates' to regulate the flow of information and better capture long-range dependencies within the sequence (Hochreiter & Schmidhuber, 1997).

After processing the entire input sentence, the final hidden state of the encoder, h_T, becomes the **context vector, C**. This vector is a fixed-size numerical summary that attempts to encapsulate the meaning of the entire source sentence. The decoder then uses this single vector as its starting point to generate the translated sentence. This architecture, however, creates an **information bottleneck**, as it forces the model to compress the meaning of potentially very long and complex sentences into one fixed-size vector, inevitably leading to information loss. This limitation was the primary motivation for developing the attention mechanism (Cho et al., 2014).

11.3 The Bidirectional RNN with Attention: A Dynamic Approach

To address the limitations of the fixed-context vector, Bahdanau et al. (2015) proposed a novel architecture that allows the decoder to "look back" at the entire source sentence at each step of the translation process. This is achieved by combining a **bidirectional RNN (BiRNN)** encoder with an **attention** mechanism.

11.3.1 Bidirectional Encoder

Instead of processing the input sentence in only one direction, a BiRNN processes it in two: a **forward pass** from start to end, and a **backward pass** from end to start (Schuster & Paliwal, 1997). This allows the hidden state for each word to contain information about both the preceding words (from the forward pass) and the following words (from the backward pass). For example, in the sentence "The man from France **speaks** fluent French," the meaning of 'speaks' is ambiguous until we see 'French.' A standard RNN would process 'speaks' without this future context. A BiRNN, however, would have its backward pass inform the representation of 'speaks' with the knowledge that "French" is coming, creating a much richer and less ambiguous representation. This process produces a sequence of annotations, h_1, h_2, ..., h_T, where each annotation, h_t, contains contextual information about the word x_t from both before and after it in the sentence. This rich sequence of annotations, rather than a single fixed vector, is then passed to the decoder (Bahdanau et al., 2015).

Figure 11.1

BiRNN Forward and Backward Pass

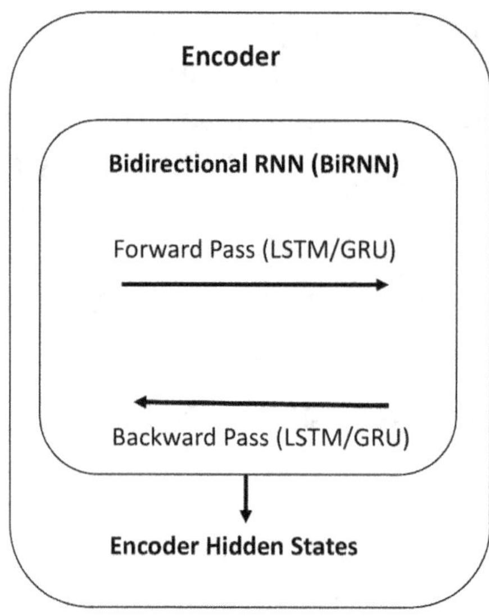

11.3.2 The Attention Mechanism

The attention mechanism is the crucial component that allows the decoder to sift through the sequence of encoder annotations and dynamically focus on the most relevant ones at each step of the generation process. It works in three steps (Bahdanau et al., 2015):

1. **Alignment Score (e_{ij}):** At each decoding step i, the decoder's current hidden state is compared with each of the encoder's hidden states, h_j, to generate an "alignment score." This score indicates how well the input word at position j aligns with the output word being generated at position i.

2. **Attention Weights (α_{ij}):** The alignment scores are passed through a **softmax function**. The purpose of the softmax function is to take a vector of arbitrary real-valued scores and transform them into a probability distribution—a set of positive numbers that sum to 1. These normalized scores become the attention weights, determining how much focus to place on each source word (Goodfellow et al., 2016).

3. **Context Vector (c_i):** The final context vector for this decoding step, c_i, is calculated as the weighted sum of all the encoder annotations.

This dynamic context vector is then used by the decoder, along with its hidden state and the previously generated word, to predict the next output word (Bahdanau et al., 2015).

Attention in Action: Example Translations

The power of attention is most evident when translating between languages with different word orders.

- Example 1: English to Spanish
  When translating "My black cat" to "Mi gato negro," the attention mechanism learns to reorder the adjective and noun.

Table 11.1

English to Spanish Example

Decoding Step	Source Word Driving Translation (Reordering Focus)	Generated Word
Step 1	*My*	*Mi*
Step 2	*cat*	*gato*
Step 3	*black*	*negro*

- Example 2: English to Japanese
  Translating "I am reading a book" to Japanese ("Watashi wa hon o yonde imasu") requires significant reordering, as Japanese is a Subject-Object-Verb language.

Table 11.2

English to Japanese Example

Decoding Step	Source Word Driving Translation (Reordering Focus)	Generated Word (Japanese)	Explanation of Attention
Step 1	*I*	私は (Watashi wa)	Model focuses on the subject *I*.
Step 2	*book*	本 (hon)	Model aligns *book* to *hon*. Japanese word order differs.
Step 3	*reading*	読んでいます (yonde imasu)	Model focuses on the verb *reading*.

Figure 11.2

The 3-Step Attention Calculation

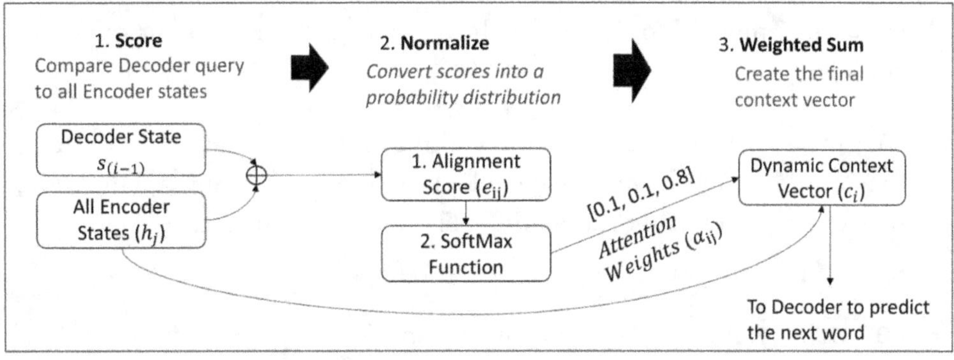

Technical Deep Dive: Calculating Attention

The attention mechanism's power is derived from three core, mathematically defined steps that allow the decoder to compute a highly relevant, dynamic context vector: the alignment score e_{ij}, the normalized attention weights a_{ij}, and the final context vector c_i. The purpose of this technical deep dive is to formally present the equations governing these steps, which form the heart of the attention model.

Alignment Score e_{ij}

The alignment score e_{ij} is calculated using a small feed-forward neural network that is learned jointly with the rest of the model. A common formulation is:

Equation 11.2

The Alignment Score (Bahdanau Attention)

$$e_{ij} = v_a^T \tanh(W_a s_{i-1} + U_a h_j)$$

- e_{ij}: The **alignment score**. This is a single scalar value that answers the question: "When generating the **i-th** word of the output, how relevant is the **j-th** word of the input?"

- s_{i-1}: The **previous hidden state of the decoder** at step i-1. It represents the summary of the output sequence generated so far. We use the *previous* state to help predict the *current* output at step i.

- h_j: The **j-th hidden state of the encoder**. This vector contains information about the j-th word (or token) of the input sequence and its surrounding context.

- W_a and U_a: These are **learnable weight matrices**. Their job is to project the decoder state s_{i-1} and the encoder state h_j into a common dimension, so they can be combined meaningfully. These matrices are learned during the training process.

- **tanh (...)**: A non-linear **activation function** (hyperbolic tangent). It takes the combined representation $(W_a s_{i-1} + U_a h_j)$ and squashes it into a vector where each element is between -1 and 1. This adds expressive power to the model.

- v_a^T : A **learnable weight vector** (the T denotes the transpose). Its purpose is to take the vector output from the tanh function and project it down to a single scalar number, which becomes our final score e_{ij}.

Attention Weights α_{ij}

The attention weights α_{ij} are then computed by applying a softmax function to these scores:

Equation 11.3

The Attention Weight Calculation (Softmax)

$$\alpha_{ij} = \frac{\exp(e_{ij})}{\sum_{k=1}^{T} \exp(e_{ik})}$$

The purpose of this formula is to convert the raw, uninterpretable alignment scores into a set of positive weights that sum to 1. This makes them behave like a probability distribution, indicating where the model should 'focus' its attention.

- α_{ij}: The **attention weight** (alpha). This is a scalar value between 0 and 1. It represents the 'importance' or 'probability' assigned to the j-th input word when generating the i-th output word. All the α_{ij} for a given output step i will sum to 1.

- **exp(...)**: The **exponential function** (e^x). This is used for two key reasons:

    - **Non-negativity**: It ensures all resulting values are positive, since e^x is always positive. Weights cannot be negative.

    - **Accentuates differences**: It exaggerates the differences between scores. A slightly higher score becomes significantly more important after the exponential function is applied, forcing the model to focus more sharply on the most relevant inputs.

- $\sum_{k=1}^{T} \exp(e_{ik})$: This is the **normalization term** (the denominator).

    - Σ: The summation symbol.

    - k=1 to T: This means we are summing over all the encoder hidden states, from the first (k=1) to the last (k=T, where T is the length of the input sequence).

    - **In plain English**: This term calculates the sum of the exponentiated scores for the current decoder step i across *all* possible input steps k.

Context Vector C_i

Finally, the context vector c_i is the weighted sum of the encoder annotations (*h*):

Equation 11.4

The Context Vector Calculation

$$c_i = \sum_{j=1}^{T} \alpha_{ij} h_j$$

This formula synthesizes the information from the entire input sequence into a single, dynamic vector that is tailored for the current output step.

- c_i: The context vector for the decoder at time step i. This vector is a summary of the input sequence, but it's a smart summary that focuses on the parts most relevant for generating the output at this specific step. It has the same dimension as the encoder hidden states.

- $\sum_{j=1}^{T}$: The summation over all the input time steps.
  - j=1 to T: This means we iterate through every hidden state from the encoder, where T is the total length of the input sequence.

- α_{ij}: The attention weight assigned to the j-th encoder hidden state for the i-th decoder step. This is the output from the softmax function you generated previously. It's a scalar value between 0 and 1 that represents the "amount of focus" on the j-th input word.

- h_j: The j-th hidden state from the encoder. This vector represents the j-th input word and its surrounding context.

By allowing the model to dynamically create a tailored context vector for each step, the attention mechanism effectively frees the NMT system from the constraints of a fixed-size representation.

11.4 Experimental Validation and Performance Analysis

To validate their approach, Bahdanau et al. (2015) conducted a comparative experiment between two NMT models:

- RNNencdec: A conventional RNN encoder-decoder with 1000 hidden units.

- RNNsearch: The proposed bidirectional RNN encoder-decoder with an attention mechanism and 1000 hidden units.

Both models were trained using minibatch stochastic gradient descent (SGD) with adadelta. Performance was evaluated using the bilingual evaluation understudy (BLEU) score, a standard metric that measures the N-gram precision overlap between a machine-generated translation and one or more high-quality human translations (Papineni et al., 2002).

The results were decisive. The RNNsearch model consistently outperformed the

RNNencdec model. Crucially, while the performance of the RNNencdec model degraded as sentence length increased, the RNNsearch model's performance remained strong, validating the hypothesis that the attention mechanism effectively overcomes the limitations of a fixed-context vector. Furthermore, the RNNsearch model achieved a BLEU score comparable to the state-of-the-art phrase-based statistical machine translation system of the time. This quantitative improvement in the BLEU score translated to significant qualitative gains: the attention-based model produced more fluent translations, handled grammatical agreement more accurately, and was far superior at translating long sentences and reordering phrases between languages with different syntactic structures (Bahdanau et al., 2015).

Figure 11.3

Attention Mechanism: Performance on Long Sentences

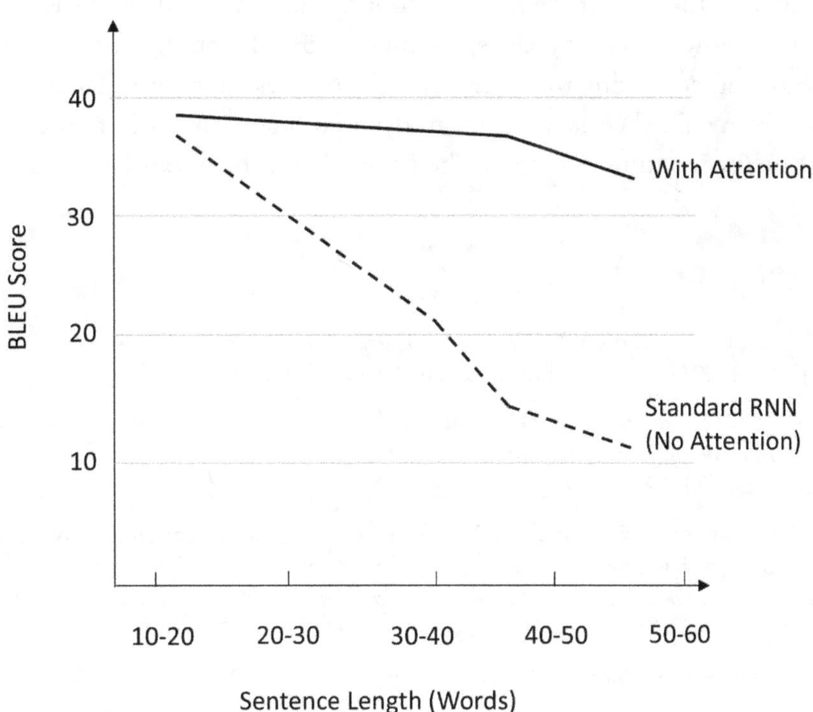

11.5 Conclusion

The work of Bahdanau et al. (2015) was a landmark achievement in the history of neural machine translation. By introducing a dynamic attention mechanism, they effectively solved the information bottleneck problem that constrained earlier RNN encoder-decoder models. The ability to create a context-aware vector at each step of the decoding process, tailored to the specific word being generated, led to a dramatic improvement in translation quality, particularly for long sentences and languages with different word orders.

This innovation did more than just improve machine translation; it introduced a powerful new concept that would become a cornerstone of modern sequence-to-sequence modeling. The concept of attention, for instance, evolved into the core component of the transformer architecture, which has since surpassed RNN-based models to become the state-of-the-art in NMT and many other NLP tasks. Future research continues to build on these foundations, exploring challenges such as translation for low-resource languages, multimodal translation that incorporates visual context, and enhancing the interpretability of these complex neural models. The journey from a fixed context vector to dynamic attention marks a true milestone, one that continues to shape the future of machine translation.

11.6 References

Bahdanau, D., Cho, K., & Bengio, Y. (2015). *Neural machine translation by jointly learning to align and translate* (arXiv:1409.0473). arXiv. https://doi.org/10.48550/arXiv.1409.0473

Bengio, Y., Simard, P., & Frasconi, P. (1994). Learning long-term dependencies with gradient descent is difficult. *IEEE Transactions on Neural Networks, 5*(2), 157–166. https://doi.org/10.1109/72.279181

Cho, K., Van Merriënboer, B., Gulcehre, C., Bahdanau, D., Bougares, F., Schwenk, H., & Bengio, Y. (2014). Learning phrase representations using RNN encoder-decoder for statistical machine translation. *In Proceedings of the 2014 Conference on Empirical Methods in Natural Language Processing (EMNLP)* (pp. 1724–1734). https://doi.org/10.3115/v1/D14-1179

Goodfellow, I., Bengio, Y., & Courville, A. (2016). *Deep learning.* MIT Press.

Hochreiter, S., & Schmidhuber, J. (1997). Long short-term memory. *Neural Computation, 9*(8), 1735–1780. https://doi.org/10.1162/neco.1997.9.8.1735

Papineni, K., Roukos, S., Ward, T., & Zhu, W. J. (2002). BLEU: A method for automatic evaluation of machine translation. *In Proceedings of the 40th Annual Meeting of the Association for Computational Linguistics* (pp. 311–318). https://doi.org/10.3115/1073083.1073135

Schuster, M., & Paliwal, K. K. (1997). Bidirectional recurrent neural networks. *IEEE Transactions on*

Signal Processing, 45(11), 2673–2681. https://doi.org/10.1109/78.650093

11.7 Glossary

Visual Glossary

Components of an Attention-Based NMT Model.

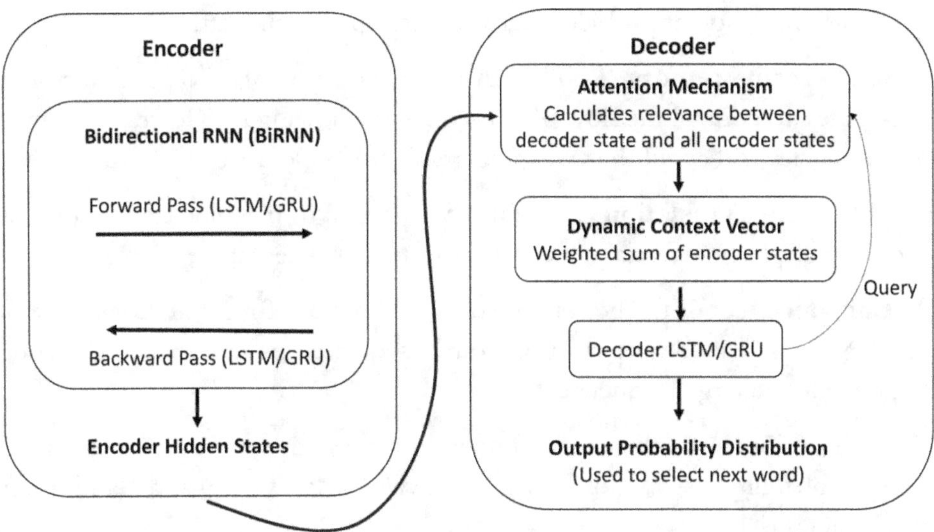

Activation Function: A mathematical function (e.g., tanh or ReLU) applied to a node in a neural network that determines its output. Non-linear activation functions allow the network to learn complex patterns in the data.

Adadelta: A specific, adaptive learning rate optimization algorithm and an advanced variant of SGD. It automatically adjusts the learning rate for each parameter during training based on past gradients, addressing the limitation of traditional SGD which requires a manually set, fixed learning rate.

Attention Mechanism: A technique that enables a model to dynamically focus on specific parts of the input sequence when producing an output, overcoming the bottleneck of a fixed-size context vector.

Bidirectional RNN (BiRNN): An enhanced RNN that processes sequence data in both forward and backward directions, allowing the representation for each token to incorporate context from both its past and future.

BLEU (Bilingual Evaluation Understudy) Score: A metric for evaluating the quality of machine-translated text by comparing n-gram precision overlap with reference translations.

Context Vector: A numerical representation (a vector) that serves as a summary of the encoder's input. In classic Seq2Seq, it is a single fixed-size vector; in attention-based models, it is a dynamic vector computed at each decoding step.

Gated Recurrent Unit (GRU): A type of recurrent neural network that, like an LSTM, uses gating mechanisms to control the flow of information and combat the vanishing gradient problem, but with a simpler architecture.

Long Short-Term Memory (LSTM): An advanced type of RNN architecture that uses 'gates' to control the flow of information, allowing it to learn long-range dependencies and avoid the vanishing gradient problem.

Neural Machine Translation (NMT): An approach to machine translation that uses a single, large artificial neural network to translate text.

RNN Encoder-Decoder: The foundational architecture for NMT, consisting of one RNN (the encoder) to read the source sequence and another (the decoder) to generate the target sequence.

Softmax Function: A mathematical function that converts a vector of real numbers into a probability distribution, where each value is between 0 and 1 and the sum of all values is 1.

Stochastic Gradient Descent (SGD): An iterative optimization algorithm used to train neural networks. It operates by calculating the gradient of the loss function over small, randomly selected subsets of data called mini-batches. This random sampling allows for faster training and efficient use of resources compared to processing the entire dataset at once.

Vanishing Gradient Problem: A difficulty encountered when training deep neural networks, where gradients become exponentially small as they are propagated back through time, preventing the network from learning long-range dependencies.

11.8 Review Questions and Discussion Topics

1. Explain in your own words the primary architectural limitation of the standard RNN encoder-decoder model. Why is this "information bottleneck" particularly problematic for long sentences?

2. What specific advantage does a bidirectional RNN encoder offer over a standard, unidirectional one? How does processing the sentence in both directions help create a richer representation of each word?

3. Describe the role of the three main steps of the attention mechanism (alignment score, attention weights, and context vector). How does this mechanism allow the model to "pay attention" to different words at different stages of the translation process?

4. The Conclusion mentions that the concept of attention evolved into the transformer architecture. Research the transformer model and discuss the key differences between its "self-attention" mechanism and the attention mechanism described in this chapter.

5. NMT systems are now widely deployed in commercial products. Discuss one potential ethical issue that could arise from deploying a large-scale NMT system (e.g., related to bias, fairness, or the spread of misinformation).

Chapter 12

A Review of Google's 2016 Neural Machine Translation System

12.1 Introduction

The previous chapter detailed the theoretical breakthrough that solved the information bottleneck in **neural machine translation (NMT)**: the dynamic attention mechanism. This concept proved that neural models could surpass Phrase-Based SMT in translation quality, especially for long sentences.

This chapter serves as the crucial bridge from that theory to production reality, exploring how the powerful idea of attention was scaled for a global consumer product. The **Google Neural Machine Translation (GNMT)** system, launched in 2016, became the definitive case study of this new paradigm. It demonstrated that a deep, complex, and highly engineered version of the attention-based architecture could achieve unprecedented fluency and accuracy, marking a milestone for the use of large-scale deep learning in a live, global service.

The purpose of reviewing GNMT is not to repeat the concept of attention, but to analyze the engineering required to deploy it. The system's design highlights innovations essential for massive scale, such as model **parallelization, quantization**, and the need to run recurrent models eight layers deep. Ultimately, this engineering effort revealed the final, unavoidable **computational bottlenecks** of recurrence (sequential processing), which directly set the stage for the breakthrough of the transformer architecture (Chapter 13).

12.2 The GNMT Architecture: An Overview

The GNMT system is a sequence-to-sequence model composed of three primary components: an encoder network, a decoder network, and an attention mechanism that links them. As illustrated in Figure 12.1, the architecture is characterized by its depth, with a stack of 8 LSTM layers for both the encoder and the decoder. This depth was a key innovation, allowing the model to learn a richer hierarchy of representations; lower layers could focus on capturing basic syntax, while higher layers could build on these to understand more abstract, semantic meaning (Wu et al., 2016).

Figure 12.1

Model Architecture of GNMT

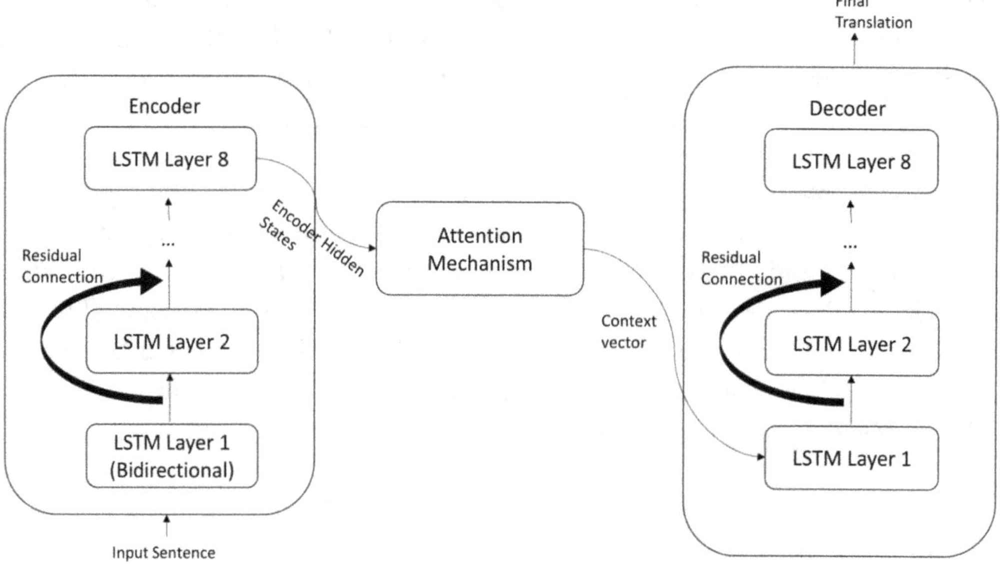

Note: The model comprises a deep LSTM encoder (left) and decoder (right), connected by an attention module. Residual connections are shown as arrows that skip over layers. Source: Wu et al. (2016).

To manage the immense computational load of such a deep model, the GNMT design incorporated sophisticated engineering solutions. **Model parallelization** was used to distribute the workload across multiple GPUs, with different layers of a single large model running on different processors. This technique **solved the**

memory challenge of deep models, but crucially, it did not eliminate the underlying sequential computational constraint inherent in the RNN architecture itself. This was a critical solution to make training the massive model feasible. Furthermore, **quantization**—reducing the precision of the model's weights (e.g., from 32-bit to 8-bit numbers)—was used to significantly speed up inference time, which was essential for deploying the model in a real-time production environment. Finally, residual connections were also employed between LSTM layers to facilitate the training of this deep network, thereby mitigating the vanishing gradient problem (He et al., 2016; Wu et al., 2016).

12.3 Core Components of the GNMT Model

The effectiveness of GNMT stems from the specialized roles of its three main architectural components, as detailed in the original paper by Wu et al. (2016).

12.3.1 Encoder Network

The role of the encoder is to read the source sentence and compress its meaning into a sequence of vector representations.

- **Structure:** The encoder consists of a multi-layer LSTM network. The first layer is bidirectional, meaning it processes the input sequence in both forward and reverse directions. This allows it to create an initial representation of each word that incorporates context from both before and after it. The subsequent seven layers are unidirectional (Wu et al., 2016).
- **Function:** The encoder takes a sequence of input words (tokens) from the source language and converts each one into a fixed-size vector. The final output of the encoder is a sequence of these vectors, one for each input word, which collectively represent the meaning of the entire source sentence (Wu et al., 2016).

12.3.2 Attention Mechanism

The attention mechanism is the bridge between the encoder and the decoder. It solves the information bottleneck of earlier Seq2Seq models by allowing the decoder to "look back" at the encoder's complete sequence of hidden states and selectively focus on the most relevant ones at each step of the translation process (Bahdanau et al., 2015).

The process, as implemented in GNMT, works as follows (Wu et al., 2016):

1. At each step in generating the target sentence, the decoder's current hidden state is compared against all the encoder's hidden states $(h_1, ..., h_T)$.

2. This comparison yields a set of "attention scores," which are normalized into probabilities (weights) via a softmax function. A high weight indicates that a particular source word is highly relevant to the target word being generated.

3. A weighted average of the encoder hidden states (the "context vector") is calculated.

4. This context vector is then fed into the decoder, providing it with precise, relevant information from the source sentence needed to generate the next target word.

12.3.3 Decoder Network

The decoder's job is to generate the translated sentence in the target language.

- **Structure:** Like the encoder, the decoder is a deep stack of eight LSTMs (Wu et al., 2016).
- **Function:** The decoder is autoregressive. At each time step, it takes the context vector provided by the attention mechanism, its own hidden state, and the word it previously generated as input. It then produces a probability distribution over the entire target vocabulary, and the word with the highest probability is chosen as the next word in the translated sentence. This process continues until an end-of-sequence token is generated (Wu et al., 2016).

12.4 Discussion: The Legacy of GNMT and the Rise of the Transformer

The introduction of GNMT in 2016 marked a watershed moment, demonstrating that deep, recurrent neural networks could achieve human-level performance on a complex NLP task at a global scale. Its success validated the power of the encoder-decoder framework augmented with the attention mechanism (Wu et al., 2016). However, the very architecture that made GNMT successful also contained inherent limitations that would inspire the next great leap forward.

The primary limitation of GNMT was its reliance on LSTMs, a type of recurrent neural network (RNN). By their nature, RNNs process data sequentially. To calculate the hidden state for a word, one must first have calculated it for the previous word. This sequential dependency creates a computational bottleneck, severely limiting the potential for parallelization during training (Jurafsky & Martin, 2023). While GNMT's engineers cleverly parallelized the model across layers on different GPUs, they could not parallelize the computations *within* each layer over time. This fundamental constraint on training speed and efficiency meant that there was a practical limit to how large and powerful these recurrent models could become.

Figure 12.2

The Sequential Bottleneck: Why RNNs Were Replaced

Then: RNN/LSTM

(Sequential Processing)

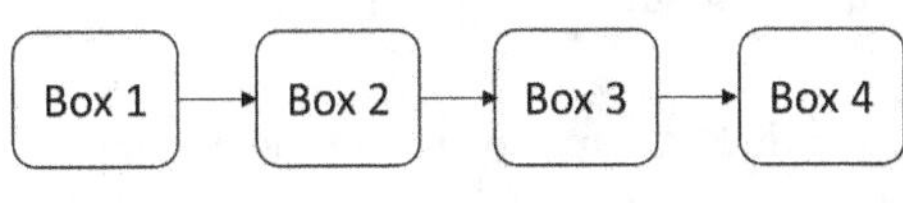

Limitations: Computation is sequential. The processing of Box 4 cannot begin until Box 3 is finished. This creates a computational bottleneck and is slow to train.

Now: Transformer

(Parallel Processing)

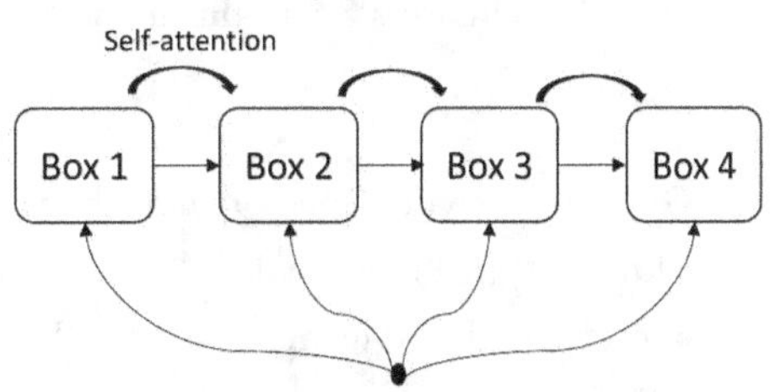

Advantage: Computation is parallel. The processing of all four boxes happens simultaneously. This is dramatically faster and more scalable, enabling the training of massive models.

12.5 Conclusion

Google's 2016 neural machine translation system was a monumental achievement in the history of artificial intelligence. Its deep, multi-layer LSTM architecture, combined with residual connections and a powerful attention mechanism, set a new standard for translation quality and demonstrated the viability of large-scale deep learning for real-world NLP applications (Wu et al., 2016).

Table 12.1.

Model Parallelization vs. Computation Parallelization

While the GNMT architecture has since been surpassed by the more efficient and

Model Parallelization (GNMT / LSTM)	Computation Parallelization (Transformer / Self-Attention)
Distribute the physical architecture of a single, massive model across multiple devices (GPUs) during training or inference.	This refers to processing the data itself simultaneously, rather than sequentially.
• **What is Parallel**: The layers of the neural network.	• **What is Parallel**: The tokens (words) in the input sequence.
• **Mechanism:** In GNMT, engineers broke the eight-layer LSTM stack vertically. For instance, LSTM Layers 1-4 might run on GPU 1, and Layers 5-8 might run on GPU 2.	• **Mechanism**: The Self-Attention mechanism in the Transformer allows the model to compute the contextualized representation for all words in a sentence at the same time.
• **Goal:** To overcome memory limitations and allow a single, extremely deep model to fit into the combined memory of several GPUs.	• **Goal:** To overcome the computational bottleneck caused by sequential processing, drastically speeding up training time.
• **Sequential Bottleneck:** This method does not change the inherent nature of the LSTM. The output of Layer 4 must still be calculated before Layer 5 can begin, and within each layer, the processing of the input sentence remains sequential (word by word).	

powerful transformer models, its legacy endures. It served as the crucial bridge between earlier, simpler sequence-to-sequence models and the modern era of attention-based architectures. By studying GNMT, we not only appreciate a landmark engineering accomplishment but also gain a clearer understanding of the evolutionary pressure created by its success. The system proved the power of attention but also highlighted the inherent computational bottlenecks of recurrence.

This created the perfect motivation for the research community to ask, "Can we build a model that keeps the power of attention while getting rid of recurrence entirely?" The answer to that question, as we will see in the next chapter, was the transformer (Vaswani et al., 2017).

12.6 References

Bahdanau, D., Cho, K., & Bengio, Y. (2015). *Neural machine translation by jointly learning to align and translate* (arXiv:1409.0473). arXiv. https://doi.org/10.48550/arXiv.1409.0473

He, K., Zhang, X., Ren, S., & Sun, J. (2016). Deep residual learning for image recognition. In *Proceedings of the IEEE Conference on Computer Vision and Pattern Recognition* (pp. 770-778). https://doi.org/10.1109/CVPR.2016.90

Jurafsky, D., & Martin, J. H. (2023). *Speech and language processing* (3rd ed.). Prentice Hall.

Vaswani, A., Shazeer, N., Parmar, N., Uszkoreit, J., Jones, L., Gomez, A. N., ... & Polosukhin, I. (2017). Attention is all you need. *Advances in Neural Information Processing Systems, 30*.

Wu, Y., Schuster, M., Chen, Z., Le, Q. V., Norouzi, M., Macherey, W., Krikun, M., Cao, Y., Gao, Q., Macherey, K., Klingner, J., Shah, A., Johnson, M., Liu, X., Kaiser, Ł., Gouws, S., Kato, Y., Kudo, T., Kazawa, H., ... Dean, J. (2016). Google's neural machine translation system: Bridging the gap between human and machine translation. https://doi.org/10.48550/arxiv.1609.08144

12.7 Glossary

Attention Mechanism: A technique that allows the decoder in a sequence-to-sequence model to dynamically focus on different parts of the input sequence when generating each token of the output.

Autoregressive: A model property where the prediction for the current time step is generated based on the outputs from previous time steps.

Decoder: The component of a sequence-to-sequence model that takes a representation of the input sequence and generates the output sequence.

Encoder: The component of a sequence-to-sequence model that reads the input sequence and compresses its information into a numerical representation.

LSTM (Long Short-Term Memory): An advanced type of recurrent neural network (RNN) that uses 'gates' to control the flow of information, allowing it to learn long-range dependencies and avoid the vanishing gradient problem.

Model Parallelization: An engineering technique for training very large neural networks where different layers of a single model are placed on different processors (e.g., GPUs) to be computed in parallel.

Quantization: A technique to reduce the computational and memory costs of running inference by representing weights and activations with lower precision data types, such as 8-bit integers instead of 32-bit floating-point numbers.

Residual Connections (or Skip Connections): A technique used in deep neural networks where the output of a layer is added to the output of a later layer. This helps gradients flow more easily through the network during training, enabling the training of much deeper models.

Self-Attention: A type of attention mechanism where the elements of a single sequence attend to each other to compute a representation of the sequence. This is the core mechanism of the transformer.

Sequence-to-Sequence (Seq2Seq): A class of models that use an encoder-decoder structure to transform an input sequence into an output sequence, where the lengths can be different.

12.8 Review Questions and Discussion Topics

1. **Role of Bidirectionality:** The GNMT encoder uses a bidirectional LSTM only for its first layer. Explain the rationale behind this design choice. What are the

benefits of making the first layer bidirectional, and why might the designers have chosen to keep the subsequent seven layers unidirectional?

2. **Attention vs. Self-Attention:** Contrast the "attention mechanism" used in GNMT with the "self-attention" mechanism that is central to the transformer architecture. How do their inputs and computational objectives differ?

3. **The Necessity of Residual Connections:** This chapter mentions that residual connections were used to facilitate the training of GNMT's deep network. Elaborate on why a deep 8-layer LSTM network might fail to train effectively without them. How, specifically, do residual connections address the vanishing gradient problem?

4. **Beyond Translation:** The sequence-to-sequence architecture is highly versatile. Describe another NLP task, besides machine translation, that could be framed as a sequence-to-sequence problem and could benefit from a deep, attention-based architecture like GNMT.

5. **Engineering for Scale:** The GNMT paper was as much an engineering achievement as a research one. Discuss the different purposes of model parallelization and quantization. How do these two techniques address the distinct challenges of training a massive model versus deploying it for real-time inference?

Chapter 13

The Transformer Revolution: Attention Is All You Need

13.1 Introduction: A Paradigm Shift from Recurrence to Attention

The field of natural language processing (NLP) underwent a profound transformation with the introduction of the transformer architecture in the 2017 paper, "Attention Is All You Need" by Vaswani et al. Prior to this, the dominant models for sequence transduction tasks, such as machine translation, were based on complex recurrent neural networks (RNNs) like LSTMs. These models process data sequentially, one token at a time, maintaining a hidden state that carries information from previous steps to subsequent ones. This sequential nature, while intuitive for modeling language, imposed a fundamental and severe limitation: the bottleneck of sequential computation. Because an RNN must process a sentence word by word, it cannot be fully parallelized, creating a hard ceiling on training speed and efficiency.

The pursuit of ever-larger and more powerful models necessitated a radical departure from this recurrent philosophy. The transformer architecture provided exactly that. It is a network architecture based solely on **attention mechanisms**, dispensing with recurrence and convolutions entirely. This design enables significantly more parallelization, allowing training on much larger datasets than previously feasible. By demonstrating that a model without recurrence could not only match but also significantly outperform recurrent models, the transformer didn't just improve upon the state of the art; it established a new foundation for nearly all modern NLP.

This chapter provides a detailed overview of the transformer architecture. We will explore its core innovations, including how it handles word order without recurrence and the elegant mechanics of self-attention. We will then walk through the full architecture, tracing the data path through the encoder and decoder stacks to build a robust mental model of how it operates. For readers interested in a deeper mathematical analysis, a comprehensive review is provided in Appendix F.

13.2 Escaping Recurrence: The Foundational Concepts

Removing recurrence creates an immediate and critical problem: the loss of sequential order. A model that processes all words at once has no inherent knowledge of which word came first. To solve this, the transformer injects positional information for each token into the input embeddings via **positional encodings**.

Think of it as giving each word an 'address.' A word's standard **embedding** tells the model *what* the word means, while its **positional encoding** tells the model *where* the word is in the sentence. These two vectors are combined to form a single vector that contains information about both meaning and position.

13.2.1 Simultaneous Input Processing in the Transformer

When we say that the transformer processes all words at once, we mean that it simultaneously generates and integrates both the word embedding and positional encoding for every token in the input sequence. The main efficiency gain starts at the input layer, where the model performs two critical functions concurrently. First, through parallel embedding generation, the model simultaneously retrieves dense vector representations (**word embeddings**) for each token from a pre-trained lookup table, capturing the meaning of each word. Second, in **positional encoding** creation, it simultaneously computes the fixed or learned vector representing each token's absolute position in the sentence for all positions.

These two vector components are then added together for every token, resulting in a single hybrid vector that contains both the semantic information ('what' the word is) and the positional information ('where' the word is). The final combined input vectors for all N tokens are then fed into the encoder stack simultaneously, which enhances the efficiency of the self-attention mechanism.

13.2.2 Enabling Parallel Computation

This simultaneous input enables the next essential step: self-attention. Because the model provides positional context upfront via positional encoding, the encoder can start calculating the contextualized representation for each word immediately, without waiting for the previous word's output. This leverages the advantages of parallel processing, as illustrated in Figure 12.2 of Chapter 12.

The foundational paper "Attention Is All You Need" utilized a clever approach involving sine and cosine waves to create these 'address' vectors. This technique assigns each position a unique mathematical fingerprint, ensuring that the relationship between positions remains consistent. As a result, the model can easily learn concepts of relative positions, such as "three words ahead" or "two words behind," which is crucial for understanding the grammatical structure of a sentence (Vaswani et al., 2017).

13.3 The Heart of the Transformer: Self-Attention

The core mechanism that allows the transformer to understand complex relationships within a sentence is **self-attention**. It allows the model to weigh the importance of all other words in the input sequence when processing a particular word (Vaswani et al., 2017).

Conceptually, for each word, self-attention creates three vectors: a **Query** (Q), a **Key** (K), and a **Value** (V). To provide a clear intuition, we can use an analogy of searching for information in a library (Jurafsky & Martin, 2023):

- **Query:** This represents the word we are currently focused on. It is like your **search query**—what you are looking for.
- **Key:** Every word in the sentence has a Key vector. It is like the **title or keyword** of a book in the library. It is used for matching.
- **Value:** Every word also has a Value vector. This is like the **actual content** of the book. It is the information we want to retrieve.

The self-attention process works as follows: for the word you are currently processing (the Query), you compare its Query vector against the Key vectors of all other words in the sentence. A high compatibility score means a word is highly relevant. These scores are then used to create a weighted sum of all the Value vectors. For example, consider the sentence: "The animal did not cross the street because it was too wide." When the model processes the word 'it' (the Query), it compares it against the Keys of 'animal' and 'street.' The model resolves the pronoun by assigning a higher attention weight to the Value of 'street' because 'wide' is a more common descriptor for a street than for an animal. This is how self-attention resolves anaphora.

The result is a new representation for your original word that is enriched with contextual information from the most relevant words in the sentence (Vaswani et al., 2017).

Table 13.1

The QKV Analogy: Focus, Index, and Content

Component	Word in Focus	Role & Analogy	Meaning in Contextualization
Query (Q)	*it*	Focus/Question: This vector asks: *What is my job right now? What am I looking for?*	The Query vector for *it* is primarily searching for a singular, non-human noun.
Key (K)	*street / animal*	Index/Label: The Key vector is a descriptor, or *label*, for every word.	The Key vectors for *street* and *animal* tell the model they are both possible singular nouns.
Value (V)	*street / animal*	Content/Information: The Value vector holds the content or definition of that word, ready to be retrieved.	The mechanism compares Q(*it*) against K(*street*) and K(*animal*). It determines the *street* is a better fit (or closer match) because *wide* is a common attribute of a street.

Table 13.2

Self-Attention Functions

Component	Type of Self-Attention	Function
Encoder	Standard Self-Attention	Allows a word to look at all other words in the input sequence to create a contextual representation. This is the most parallel operation.
Decoder	Masked Self-Attention	Allows a word to look only at the words that have already been generated in the output sequence (i.e., tokens before it), preventing it from cheating by looking ahead (See section 13.4).

To empower the model to capture a richer set of relationships simultaneously, the transformer employs **multi-head attention**. This mechanism doesn't perform attention just once; it runs the process multiple times in parallel, with each 'head' learning to focus on different linguistic aspects (e.g., one head might track syntactic dependencies, while another tracks semantic similarity). This allows the model to jointly attend to information from different representation subspaces at different positions (Vaswani et al., 2017).

13.4 Assembling the Pieces: The Full Transformer Architecture

The full transformer architecture can be understood with an analogy. Think of the encoder's job as creating a deep, contextual understanding of a sentence, and the decoder's job as using that understanding to write a new sentence.

Figure 13.1

High-Level Transformer Architecture

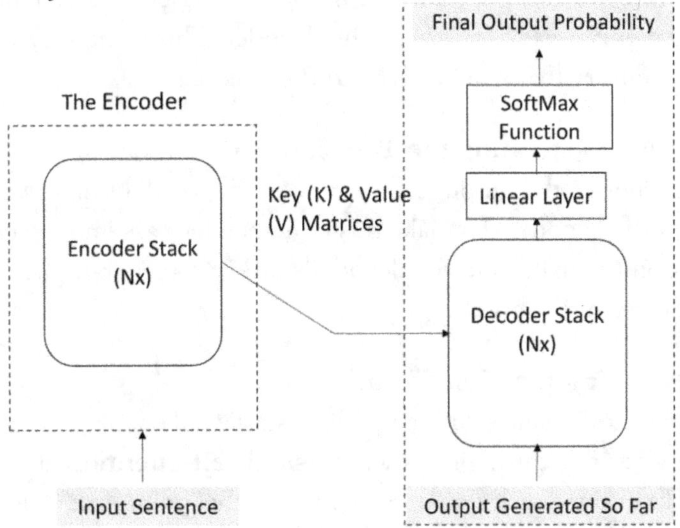

Note: This simplified diagram shows the overall flow, including the encoder producing Key (K) and Value (V) matrices used by the decoder's attention.

Part 1: The Encoder - Creating a "Rich Understanding" of the Input The encoder's goal is to take a simple sequence of words and enrich each word with context from the entire sentence.

- **Step 1: Input with 'Addresses'.** The process begins with the input embeddings combined with their positional encodings (the 'address' for each word).
- **Step 2: The Encoder Layers (The "Expert Discussion").** The vectors

for each word then pass through a stack of encoder layers. Each layer is like a round of discussion:

- **Self-Attention:** Each word's representation is updated by "looking at" all other words in the sentence. For example, in "The cat sat on the mat," the representation for 'sat' would be strongly influenced by 'cat.' This is where the model builds its deep contextual understanding.
- **Feed-Forward Network:** After the discussion, each word's representation is processed individually. This is like each expert taking a moment to think and refine their understanding.

- **Step 3: Final Encoder Output.** After passing through all the layers, the final output is a set of contextually rich vectors (one for each input word). This output can be thought of as the model's complete 'understanding' of the input sentence, ready for use by the decoder. This is the **Key (K)** and **Value (V)** memory that will be given to the decoder.

Part 2: The Decoder - Generating the Translation One Word at a Time. The decoder's job is to generate the output sentence word by word, using the rich understanding from the encoder. To make this concrete, let's assume the encoder input is "The cat is on the mat," and the decoder's task is to generate the French translation, "Le chat est sur le tapis."

- **Step 1: Generating the First Word.**
  - The decoder starts with a special <start> token.
  - It passes this through its own **masked self-attention** layer, which looks at the words generated so far (at this point, just <start>).
  - Then, in the crucial **encoder-decoder attention** step, it uses its current representation as a **Query (Q)** to 'ask' the encoder's memory (the K and V output): "Given the input 'The cat is on the mat' and that I'm starting a sentence, what part of the input should I focus on?"
  - Based on this focused information, it predicts the most probable first word of the French translation: **'Le'** (the French word for 'The').
- **Step 2: Generating the Second Word.**
  - The decoder now takes the word it just generated ('Le') as additional input.
  - The masked self-attention layer now looks at both <start> and 'Le.'
  - The encoder-decoder attention layer forms a new Query (now influenced by 'Le') and asks the encoder's memory again: "Now that

I've written 'Le', what part of the input should I focus on to continue the phrase 'The cat...'?"

  ○ It uses this new focused information to predict the second word: **'chat'** (the French word for 'cat').

- **This process repeats:** The decoder continues to generate one word at a time, feeding its own output back into itself, until it generates the full sentence and a special <end> token.

Figure 13.2

Detailed Transformer Layer Structure

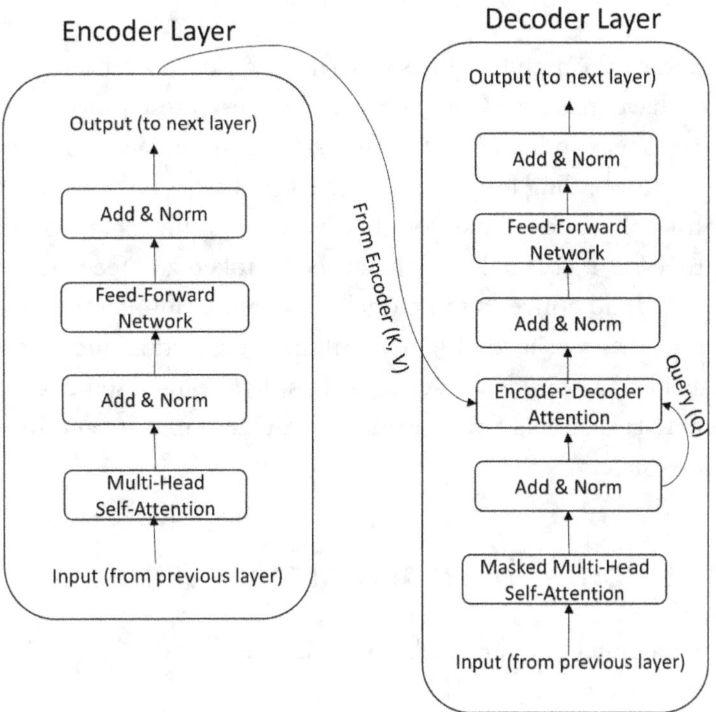

Note: The transformer's "encoder layer" is composed of two sub-layers:

- Multi-Head Self Attention
- Feed-Forward Network

The diagram is labeled "encoder stack (Nx)," indicating that the entire "encoder layer" block is stacked N times (e.g., 6 times). The output from that top 'Norm' block is the final, stabilized output of "encoder layer 1." This output is then fed as the input to the bottom of "encoder layer 2." This process repeats all the way up the stack.

The same logic applies to the transformer's "decode layer", but with three sublayers

- Masked Multi-Head Self-Attention
- Multihead Encoder-Decoder Attention
- Feed-Forward Network

As with the encoder stack (Nx), the decoder stack (Nx) repeats itself N times.

Crucially, the final output from the N-th (top) decoder layer is *not* the end. This final output is then passed to a **separate, final linear layer and SoftMax function** (as shown in Figure 13.1) to generate the actual output probability.

13.5 In Practice: Using a Transformer with Hugging Face

While understanding the architecture is crucial, modern NLP libraries like *Hugging Face's transformers* have made it remarkably simple to use these powerful pre-trained models. A practitioner can load a state-of-the-art, pre-trained sentiment analysis model and its corresponding tokenizer with a single command. The library then provides a high-level 'pipeline' function that bundles all the necessary steps: it takes raw text, preprocesses it correctly with the specific tokenizer, feeds the result through the model, and converts the model's raw output into a human-readable prediction. This allows a user to obtain a sentiment prediction, including a 'positive' or 'negative' label and a confidence score, by simply feeding a sentence into the pipeline object. This abstracts away the architectural complexity and makes state-of-the-art NLP accessible.

13.6 Discussion: Why Transformers Won

The transformer architecture quickly became the new standard in NLP for two primary reasons (Vaswani et al., 2017):

1. **The Parallelization Breakthrough:** Because the model contains no recurrence, the computations for each word in a sequence can be performed in parallel. This allows researchers to train models on vastly larger datasets in a fraction of the time it would take with an RNN.

2. **Superior Handling of Long-Range Dependencies:** In an RNN, the signal from a word must propagate through many sequential steps to reach a distant word, risking information loss due to the vanishing gradient problem (Bengio et al., 1994). In a transformer, every word has a direct path to every

other word through the self-attention mechanism, making it far more effective at capturing long-range contextual relationships.

This combination of training efficiency and modeling power enabled the training of massive, pre-trained foundational models such as BERT and the GPT series, which have revolutionized the field (Devlin et al., 2019; Brown et al., 2020). However, the architecture is not without its own limitations. The most significant is that the memory and computation required for self-attention grow quadratically with the sequence length ($O(n^2)$), making it very expensive for extremely long documents. This has spurred a new wave of research into more efficient attention mechanisms (Tay et al., 2020).

13.7 Conclusion: A New Foundation for NLP

The transformer did more than just set a new performance benchmark; it introduced a new way of thinking about sequence modeling. By demonstrating that recurrence was not a prerequisite for understanding language, it opened the door to a new class of highly parallelizable, attention-based architectures. This foundational shift is what enabled the scaling laws that have led to the massive large language models (LLMs) of today. While the quadratic complexity of attention remains a challenge that researchers are actively addressing, the core principles of the transformer—self-attention and positional encodings—have become the undisputed foundation of modern natural language processing.

13.8 References

Bengio, Y., Simard, P., & Frasconi, P. (1994). Learning long-term dependencies with gradient descent is difficult. *IEEE Transactions on Neural Networks, 5*(2), 157–166. https://doi.org/10.1109/72.279181

Brown, T. B., Mann, B., Ryder, N., Subbiah, M., Kaplan, J., Dhariwal, P., Neelakantan, A., Shyam, P., Sastry, G., Askell, A., Agarwal, S., Herbert-Voss, A., Krueger, G., Henighan, T., Child, R., Ramesh, A., Ziegler, D. M., Wu, J., Winter, C., … Amodei, D. (2020). Language models are few-shot learners. https://doi.org/10.48550/arxiv.2005.14165

Devlin, J., Chang, M. W., Lee, K., & Toutanova, K. (2019). BERT: Pre-training of deep bidirectional transformers for language understanding. *In Proceedings of the 2019 Conference of the North American Chapter of the Association for Computational Linguistics: Human Language Technologies*, Volume 1 (Long and Short Papers) (pp. 4171–4186). https://doi.org/10.18653/v1/N19-1423

Jurafsky, D., & Martin, J. H. (2023). *Speech and language processing* (3rd ed.). Prentice Hall.

Tay, Y., Dehghani, M., Bahri, D., & Metzler, D. (2020). *Efficient transformers: A survey.* arXiv preprint arXiv:2009.06732. https://arxiv.org/abs/2009.06732

Vaswani, A., Shazeer, N., Parmar, N., Uszkoreit, J., Jones, L., Gomez, A. N., ... & Polosukhin, I. (2017). Attention is all you need. *Advances in Neural Information Processing Systems, 30.*

13.9 Glossary

Decoder: The component of a transformer model responsible for generating the output sequence, using information from the encoder and the sequence it has generated so far.

Embeddings: Dense vector representations of tokens in a multi-dimensional space, where semantic similarity corresponds to proximity.

Encoder: The component of a transformer model responsible for reading the input sequence and creating a rich, contextualized representation.

Feed-Forward Network: A sub-layer within a transformer block that consists of a simple, fully connected neural network, used to process the output of the attention mechanism.

Masking: A process of hiding or ignoring certain inputs from the model to prevent it from 'seeing' information it shouldn't have access to, such as future tokens in a sequence it is trying to predict.

Multi-Head Attention: A mechanism that runs the self-attention process multiple times in parallel and concatenates the results, allowing the model to jointly attend to information from different representation subspaces.

Pipeline: A high-level helper function, commonly found in libraries like Hugging Face's transformers, that encapsulates all the steps required to perform an NLP task (such as preprocessing, model inference, and post-processing) into a single, easy-to-use object.

Positional Encoding: A vector added to the input embeddings to provide the model with information about the position of each word in the sequence, compensating for the lack of recurrence.

Query, Key, Value (QKV): A set of three vectors derived from each input embedding. The Query represents the current word's focus, the Key represents a word's 'label' for being searched, and the Value represents the word's actual content.

Recurrent Neural Network (RNN): A type of neural network designed to work with sequential data, where connections between nodes form a directed graph along a temporal sequence.

Self-Attention: The core mechanism of the transformer, which allows the model to weigh the importance of all other words in a sequence when processing any given word.

Tokenization: The process of breaking down a stream of text into smaller units called tokens, which can be words, subwords, or characters.

Transformer: A neural network architecture that relies on self-attention mechanisms instead of recurrence to process sequential data, enabling massive parallelization and superior handling of long-range dependencies.

13.10 Review Questions and Discussion Topics

1. **Core Innovation:** Explain in your own words why removing recurrence was the key innovation of the transformer. What problem did it solve, and what new problem did it create?

2. **Positional Encoding:** Why are positional encodings a necessary component of the transformer architecture? What kind of information would be lost if they were removed?

3. **Decoder Architecture:** Describe the purpose of the two different attention sub-layers within the transformer's decoder block (masked self-attention and encoder-decoder attention). Why are both necessary for the task of translation?

4. **Parallelization:** Elaborate on why an RNN/LSTM cannot be parallelized across the sequence length dimension in the same way a transformer can.

5. **Foundational Models:** Models like BERT and GPT are built on the Transformer architecture. Research either BERT or GPT and explain how it adapts the original transformer design for its specific task (e.g., language understanding vs. language generation).

PART IV

Applications and Advanced Topics

For detailed, end-to-end examples of these concepts in action, readers can refer to the practical case studies in Appendix H

Chapter 14

Text Analysis in the Age of Artificial Intelligence

14.1 Introduction: The Primacy of Language in Modern AI

The revolution in artificial intelligence (AI) is, in many ways, a revolution in understanding language. While data has long been hailed as the fuel of the digital age, it is the vast, unstructured expanse of human language—in books, on the web, in emails, and on social media—that represents both the greatest challenge and the most profound opportunity for creating truly intelligent systems. The ability to transform this raw textual data into structured, meaningful, and predictive insight is now a key driver of competitive advantage and scientific discovery. This chapter outlines the progression from the principles of structured data analysis to the advanced techniques of modern natural language processing (NLP), which underpin today's AI. We will explore the foundational methods required to make text computationally accessible, survey the core analytical tasks that extract value from it, and look ahead to a future where the interaction between humans and AI is increasingly conversational and contextual. Furthermore, for readers interested in how these analytical methods are applied in end-to-end systems, Appendix H: *Extended Practical Case Studies* provides detailed, hands-on examples for sentiment analysis, document summarization, and conversational AI.

14.2 Context and Origins: From Data Mining to Text Mining

The discipline of extracting knowledge from data, broadly known as data mining, provides the intellectual foundation for text analysis. Data mining is the process of discovering meaningful patterns in large, typically structured datasets, which are characterized by their high degree of organization in databases, spreadsheets, or enterprise systems (Han & Tong, 2022). It employs a range of supervised learning techniques (like classification and regression, where models learn from labeled data) and unsupervised learning techniques (like clustering, which finds inherent groupings in unlabeled data) to identify trends and make predictions.

However, the rigid schemas of structured data do not reflect the world of human communication. Text is inherently unstructured, ambiguous, and context-dependent. A single word can have multiple meanings, and the sentiment of a sentence can be inverted by a single turn of phrase. This necessitated the development of **text mining** (also known as text analytics), a specialized subfield that adapts the principles of data mining to the unique challenges of textual data. The central goal of text mining is to first impose structure on unstructured text and then use this structure to identify patterns and derive new knowledge (Feldman & Sanger, 2007).

14.3 The Text Transformation Pipeline: From Words to Vectors

Before any analysis can occur, raw text must be transformed into a structured, numerical format that a machine can understand. This involves a two-stage pipeline: normalization and vectorization (Aggarwal & Zhai, 2012).

- **Normalization:** As discussed in Chapter 4, this initial step cleans the text by breaking it into tokens and standardizing them through processes like stop word removal and lemmatization.
- **Vectorization- From Word Frequencies to Semantic Vectors:** Once text is normalized, it must be converted into a quantitative format. Classic approaches like TF-IDF weight words by their importance in a document (Jones, 1972; Virumeshwaran & Thirumahal, 2024). However, modern NLP is built upon **embeddings**—dense vector representations that capture semantic meaning (Mikolov et al., 2013).

The evolution from static to contextual embeddings marks one of the most significant revolutions in modern NLP. Static embeddings, such as those produced by Word2Vec (Mikolov et al., 2013) or GloVe (Pennington et al., 2014), learn a single, fixed vector representation for each word in the vocabulary. This was a massive leap forward from simple word counts, as it captured semantic relationships; for example, the vectors for 'king' and 'queen' would be located in a similar region of the vector space. However, this approach has a critical flaw: it cannot handle polysemy, the phenomenon where a single word has multiple meanings. A static model generates only one vector for the word 'bank,' which is forced to be a confusing average of its financial and geographical meanings. This limitation means the model cannot distinguish between "opening a bank account" and "sitting on a river bank." Contextual embeddings, pioneered by models like ELMo (Peters et al., 2018) and perfected by the transformer architecture in models like BERT (Devlin et al., 2019), solve this problem entirely. Instead of a fixed dictionary of word vectors, these models generate a new, unique vector for each word every time it appears, based on the specific context of the sentence it is in. The vector for 'bank' in "river bank" will be located in a region of the vector space close to words like 'water' and 'shore,' while the vector for 'bank' in "investment bank" will be close to words like 'finance' and 'money.' This ability to generate context-dependent representations was the key to unlocking a much deeper, more nuanced level of language understanding in machines.

Figure 14.1

The Evolution of Meaning: Static vs. Contextual Embeddings

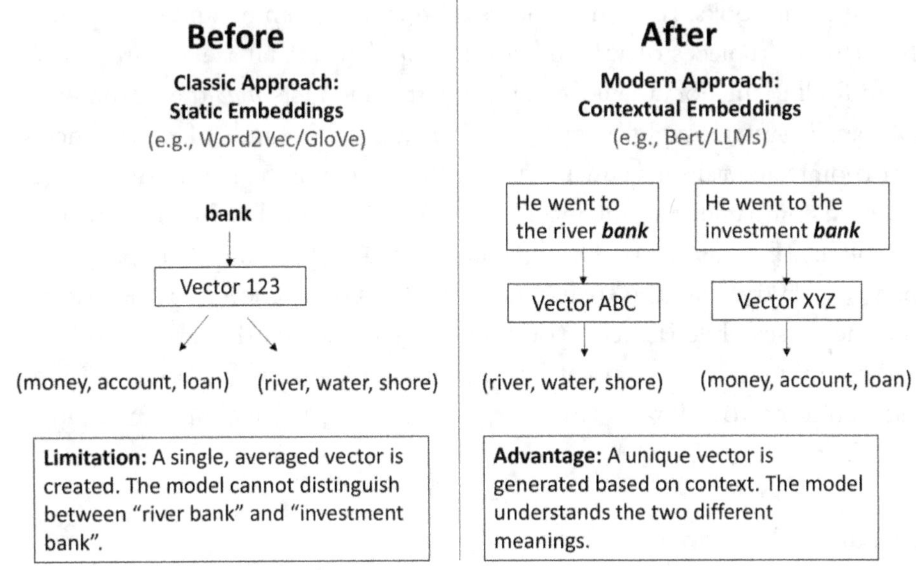

14.4 Core Text Analysis Methods in AI

With text transformed into a meaningful numerical format, a variety of analytical methods can be applied. These methods form the core capabilities of modern text analysis systems (Aggarwal, 2018).

14.4.1 Information Retrieval (IR) and Information Extraction (IE)

Information retrieval (IR) and information extraction (IE) are two foundational but distinct tasks in text analysis. While often used interchangeably, they have fundamentally different goals. Understanding their relationship is key to understanding how we transform raw text into knowledge. A useful analogy is to think of the process as working in a massive library. **Information retrieval (IR) is** the task of finding the right documents from a vast collection. It is equivalent to using the library's card catalog or search engine to find a set of relevant books based on a query. The output of an IR system is a ranked list of documents. Search engines like Google are the most prominent examples of IR systems; their goal is to retrieve the web pages most relevant to a user's search terms (Manning et al., 2008).

Information Extraction (IE), on the other hand, begins where IR ends. Once you have the relevant books, IE is the process of opening them up and pulling out specific, structured pieces of information to populate a database or spreadsheet. It is not about finding the document, but about extracting facts *from* the document. For example, an IE system might be tasked with reading thousands of news articles about corporate acquisitions and automatically extracting the names of the acquiring and acquired companies and the monetary value of the deal. While IR returns a list of documents, IE returns structured data entries like acquisition(Company_A, Company_B, $500_Million) (Grishman, 2019). This distinction is crucial: IR helps you find the noise, while IE helps you find the signal within that noise. Both are essential components of a complete text analysis pipeline, with IR often serving as the first step to narrow down a large corpus before the more computationally intensive IE task begins.

14.4.2 Topic Modeling

Topic modeling is an unsupervised learning technique for discovering abstract 'topics' in a collection of documents. It is particularly useful for getting a high-level understanding of a large, unlabeled text corpus. The most well-known algorithm for this task is **latent Dirichlet allocation (LDA)** (Terenin et al., 2019). The core intuition behind topic modeling is the idea that documents are not about a single topic but are composed of a mixture of several topics, and that each of these hidden topics can be characterized by a particular distribution of words. For example, in a corpus of news articles, one topic discovered by the model might be heavily weighted with words like 'election,' 'vote,' 'candidate,' and 'poll,' which a human would easily label as 'politics.' Another topic might be defined by words like 'market,' 'stock,' 'trade,' and 'economy,' which corresponds to 'finance.' The LDA algorithm works by iteratively processing the corpus to determine, for each word in each document, which hidden topic it is most likely to have been generated from. By repeating this process thousands of times, it converges on a stable representation of the topics (the word distributions) and the topic mixture for each document. This allows systems to automatically organize large volumes of text into coherent themes without any prior labeling, making it a powerful tool for exploring large datasets.

14.4.3 Text Classification and Clustering

Text classification is a supervised learning task that assigns predefined labels to documents. A classic example is email spam detection, where a model is trained on labeled examples to classify new emails as 'spam' or "not spam." Text clustering is its unsupervised counterpart, grouping similar documents together based on their content without any predefined categories (Aggarwal & Zhai, 2012).

14.4.4 Sentiment Analysis

Sentiment analysis, or opinion mining, is a specialized form of text classification that aims to determine the emotional tone or subjective opinion expressed in a piece of text, typically categorizing it as positive, negative, or neutral. It is one of the most commercially significant applications of NLP, used for brand monitoring, product review analysis, and gauging public opinion (Pang & Lee, 2008).

14.5 Evaluating Performance in Text Analysis

To measure the effectiveness of these methods, a suite of standard metrics is used, especially for classification and retrieval tasks. As detailed in Chapter 15, these include accuracy, precision, recall, and the F1-Score. For ranked retrieval, metrics like mean average precision (MAP) and normalized discounted cumulative gain (NDCG) are used (Manning et al., 2008).

14.6 Applications in Modern AI Systems

The methods described above are integral components of modern AI applications (Aggarwal, 2018):

- **In Academic Research:** Analyzing millions of scholarly articles to identify research gaps and emerging trends.

- **In Healthcare:** Extracting patient symptoms and outcomes from unstructured clinical notes to predict disease progression.

- **In Business Intelligence:** Analyzing customer feedback from reviews and surveys to drive product strategy.

- **In Social Media Analysis:** Tracking public opinion and brand sentiment in real-time.

14.7 The Future is Conversational: Retrieval-Augmented Generation (RAG)

Looking ahead, the future of text analysis is deeply intertwined with the capabilities of large language models (LLMs). A cutting-edge technique that exemplifies this future is **retrieval-augmented generation (RAG)** (Lewis et al., 2020). This architecture represents a powerful fusion of information retrieval and language generation, designed to make LLMs more factual, verifiable, and up-to-date. Instead of relying solely on the vast but static knowledge 'memorized' during its training, a RAG system first takes a user's prompt and uses it as a query for an information retrieval system. This system searches over a specific, trusted knowledge base—such as a company's internal technical manuals, the latest medical research papers, or a live index of the web. The top-ranked, most relevant documents are then retrieved and provided to the LLM as additional context, along with the original prompt. The LLM is then instructed to synthesize an answer based only on the provided information. This process 'augments' the model's internal knowledge with external, real-time data, grounding its response in verifiable facts and significantly reducing its tendency to 'hallucinate' or provide outdated information. RAG represents a major step towards building more reliable and trustworthy AI systems (Gao et al., 2023)

Figure 14.2

The RAG (Retrieval-Augmented Generation) Architecture

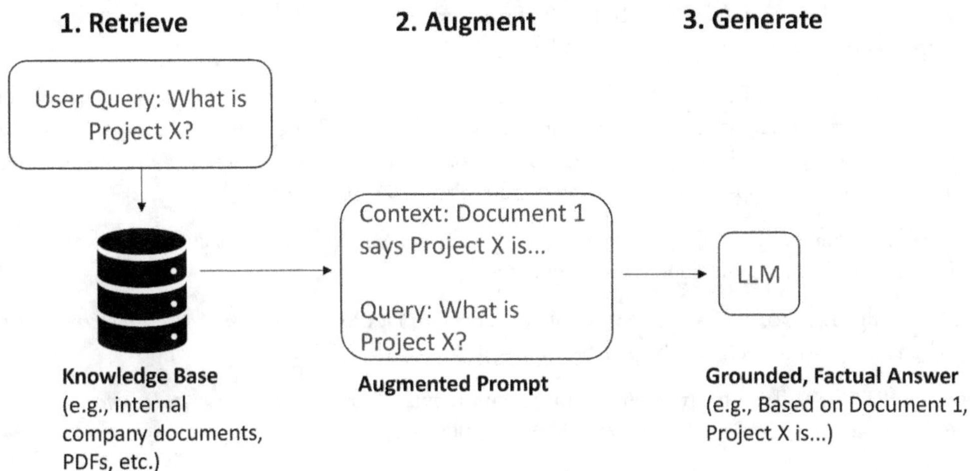

14.8 Conclusion

The journey from **data mining** to **text mining** has fundamentally reshaped how we interact with information. We have moved from analyzing structured spreadsheets to deriving meaning from the vast, messy world of human language. The evolution of text representation, from simple word counts to rich, contextual embeddings, has been the primary driver of this progress, enabling the development of powerful analytical methods that can classify, cluster, and extract information with increasing accuracy. These capabilities are no longer niche academic pursuits; they are the engine of modern AI, powering everything from search engines to business intelligence dashboards. However, significant challenges remain that will define the next era of NLP research, including the pursuit of commonsense reasoning, the mitigation of bias, and ensuring the factuality and ethical use of these powerful technologies. The journey from data mining to text mining has brought us to the forefront of AI, a quest to build machines that can understand and interact with the world in a more meaningful and responsible way.

14.9 References

Aggarwal, C. C. (2018). *Machine learning for text*. Springer. https://doi.org/10.1007/978-3-319-73531-3

Aggarwal, C. C., & Zhai, C. (2012). A survey of text classification algorithms. In C. C. Aggarwal & C. Zhai (Eds.), *Mining text data* (pp. 163–222). Springer. https://doi.org/10.1007/978-1-4614-3223-4_6

Devlin, J., Chang, M. W., Lee, K., & Toutanova, K. (2019). BERT: Pre-training of deep bidirectional transformers for language understanding. *In Proceedings of the 2019 Conference of the North American Chapter of the Association for Computational Linguistics: Human Language Technologies*, Volume 1 (Long and Short Papers) (pp. 4171–4186). https://doi.org/10.18653/v1/N19-1423

Feldman, R., & Sanger, J. (2007). The text mining handbook: Advanced approaches in analyzing unstructured data. Cambridge University Press.

Gao, Y., Xiong, Y., Gao, X., Jia, K., Pan, J., Bi, Y., ... & Sun, H. (2023). *Retrieval-augmented generation for large language models: A survey*. https://arxiv.org/abs/2312.10997

Grishman, R. (2019). Twenty-five years of information extraction. *Natural Language Engineering, 25*(6), 677-690. https://doi.org/10.1017/S1351324919000512

Han, J., Pei, J., & Tong, H. (2022). *Data mining: Concepts and techniques* (Fourth edition). Morgan Kaufmann.

Jones, K. S. (1972). A statistical interpretation of term specificity and its application in retrieval. *Journal of Documentation, 28*(1), 11–21. https://doi.org/10.1108/eb026526

Lewis, P., Perez, E., Piktus, A., Petroni, F., Karpukhin, V., Goyal, N., ... & Kiela, D. (2020). Retrieval-augmented generation for knowledge-intensive NLP tasks. *Advances in Neural Information Processing Systems, 33*, 9459–9474. https://arxiv.org/abs/2005.11401

Manning, C. D., Raghavan, P., & Schütze, H. (2008). *Introduction to information retrieval*. Cambridge University Press. https://doi.org/10.1017/CBO9780511809071

Mikolov, T., Chen, K., Corrado, G., & Dean, J. (2013). *Efficient estimation of word representations in vector space*. https://arxiv.org/abs/1301.3781

Pang, B., & Lee, L. (2008). Opinion mining and sentiment analysis. *Foundations and Trends in Information Retrieval, 2*(1–2), 1–135. https://doi.org/10.1561/1500000011

Pennington, J., Socher, R., & Manning, C. D. (2014). GloVe: Global vectors for word representation. *In Proceedings of the 2014 Conference on Empirical Methods in Natural Language Processing (EMNLP)* (pp. 1532–1543). https://doi.org/10.3115/v1/D14-1162

Peters, M. E., Neumann, M., Iyyer, M., Gardner, M., Clark, C., Lee, K., & Zettlemoyer, L. (2018). Deep contextualized word representations. *In Proceedings of the 2018 Conference of the North American Chapter of the Association for Computational Linguistics: Human Language Technologies*, Volume 1 (Long Papers) (pp. 2227–2237). https://doi.org/10.18653/v1/N18-1202

Terenin, A., Magnusson, M., Jonsson, L., & Draper, D. (2019). Pólya urn latent Dirichlet allocation: A doubly sparse massively parallel sampler. *IEEE Transactions on Pattern Analysis and Machine Intelligence, 41*(7), 1709–1719. https://doi.org/10.1109/TPAMI.2018.2832641

Virumeshwaran, M., & Thirumahal, R. (2024). TF-IDF vectorization and clustering for extractive text summarization. *Journal of Information Technology and Digital World, 6*(1), 96–111.

https://doi.org/10.36548/jitdw.2024.1.008

14.10 Glossary

Embeddings: Dense vector representations of words or sentences in a multi-dimensional space, where semantic similarity corresponds to proximity.

Information Extraction (IE): The task of automatically extracting structured information (such as entities and relationships) from unstructured text.

Information Retrieval (IR): The field concerned with finding documents or information relevant to a user's query from a large collection.

Large Language Model (LLM): A deep learning model, typically based on the transformer architecture, trained on massive amounts of text data to understand and generate human-like language.

Retrieval-Augmented Generation (RAG): An architecture that combines a generative LLM with an external information retrieval system to ground responses in specific, retrieved data.

Sentiment Analysis: The use of NLP to identify and quantify subjective opinions and emotional tones within text. Also known as opinion mining.

Text Mining: The process of deriving high-quality information from unstructured text by transforming it into a structured format to identify meaningful patterns.

TF-IDF (Term Frequency-Inverse Document Frequency): A statistical measure used to evaluate the importance of a word in a document relative to a collection of documents.

Topic Modeling: An unsupervised machine learning technique used to discover the abstract 'topics' that occur in a collection of documents.

14.11 Review Questions and Discussion Topics

1. **Representation Matters:** Compare and contrast TF-IDF and contextual embeddings (like those from BERT) as methods for text representation. In what scenarios might a simpler model like TF-IDF still be a viable choice?

2. **Practical Application Design:** Imagine you are tasked with building an AI system for a large hospital network to analyze patient experience from thousands of unstructured text reviews. Outline a text analysis pipeline for this task, specifying the goals, preprocessing steps, core analytical methods, and evaluation metrics.

3. **The Future of Knowledge - RAG vs. Scale:** The chapter introduces retrieval-augmented generation (RAG). This approach contrasts with simply building ever-larger LLMs that 'memorize' knowledge. Discuss the trade-offs between these two approaches, considering issues of factuality, cost, and the ability to incorporate new information.

Chapter 15

A Guide to Evaluation in NLP

15.1 Introduction: The Science of Measurement

In the world of natural language processing, building a model is only half the battle. How do we know if a machine translation system is accurate? How can we prove a new sentiment classifier is better than the last? The answer lies in evaluation, a critical and nuanced discipline dedicated to assessing the performance of NLP systems. Without rigorous evaluation, progress is impossible to measure, and the comparison of different models becomes subjective and unreliable.

This chapter provides a comprehensive guide to the methodologies and metrics that form the bedrock of NLP evaluation. We will begin by distinguishing between two fundamental evaluation philosophies: intrinsic and extrinsic. We will then take a deep dive into the specific metrics used for the field's most common tasks, including classification, generation, and information retrieval. Finally, we will discuss the ultimate benchmark—human evaluation—and why, despite the utility of automatic metrics, it remains the indispensable gold standard for judging true language understanding.

15.2 Intrinsic Vs. Extrinsic Evaluation

The first step in evaluating an NLP system is to decide what to measure. This choice leads to two broad approaches (Sparck Jones & Galliers, 1996; Jurafsky & Martin, 2023):

- **Intrinsic Evaluation:** This approach measures the performance of a model on a specific, isolated subtask. The evaluation is self-contained and based on a predefined set of criteria and a ground-truth dataset. It is fast, cheap, and allows for rapid, repeatable experiments, making it the standard for academic research and model development.

- **Extrinsic Evaluation:** This approach, also known as end-to-end evaluation, measures the impact a component has on the performance of a larger, downstream application. It is slower and more expensive but provides a true measure of the model's real-world utility.

To make this distinction concrete, let's consider the development of a new syntactic parser, a model that analyzes the grammatical structure of sentences. An intrinsic evaluation would involve testing the new parser on a 'treebank,' a corpus of sentences that have been manually annotated with their correct grammatical structures by linguists. The parser's performance would be measured by how closely its output matches these "gold standard" trees, using metrics like F1-score (Manning & Schütze, 1999). This is a purely academic, self-contained test of the parser's technical accuracy. However, this high intrinsic score doesn't guarantee real-world value.

An extrinsic evaluation, on the other hand, would integrate this new parser into a real-world application, such as a customer service chatbot. The goal of the chatbot is to understand user requests and route them to the correct department. The evaluation would not measure the parser's accuracy directly. Instead, it would measure key business metrics: Did the chatbot with the new parser successfully resolve more user queries? Did the rate of users escalating to a human agent decrease? Did customer satisfaction scores improve? This extrinsic evaluation measures the parser's true impact on a downstream task. A parser with a slightly lower intrinsic score might actually perform better extrinsically if its specific error patterns are less disruptive to the chatbot's logic. For example, a chatbot's logic may only need to extract a key verb (e.g., 'book') and a key entity (e.g., 'Boston'). A parser that makes a minor grammatical error (a "non-disruptive" error) but successfully extracts 'book' and 'Boston' is extrinsically more valuable than a high-scoring 'academic' parser whose errors catastrophically fail to identify the main intent (a 'disruptive' error). This highlights the critical trade-off: intrinsic evaluations are essential for rapid scientific progress, but extrinsic evaluations are necessary to prove real-world value (Jurafsky & Martin, 2023).

15.3 Metrics for Classification Tasks

For classification tasks like sentiment analysis or topic labeling, evaluation begins with a **confusion matrix**. This table breaks down a model's predictions into four categories (Fawcett, 2006):

- **True Positives (TP):** The model correctly predicted the positive class (e.g., correctly identified a spam email as spam).
- **True Negatives (TN):** The model correctly predicted the negative class (e.g., correctly identified a legitimate email as not spam).
- **False Positives (FP):** The model incorrectly predicted the positive class (e.g., misidentified a legitimate email as spam). This is also known as a **type I error.**
- **False Negatives (FN):** The model incorrectly predicted the negative class (e.g., missed a spam email and classified it as not spam). This is also known as a **type II error.**

15.3.1 The Accuracy Paradox

Figure 15.1

The Accuracy Paradox: Why 99.9% Accuracy Can Be a Complete Failure

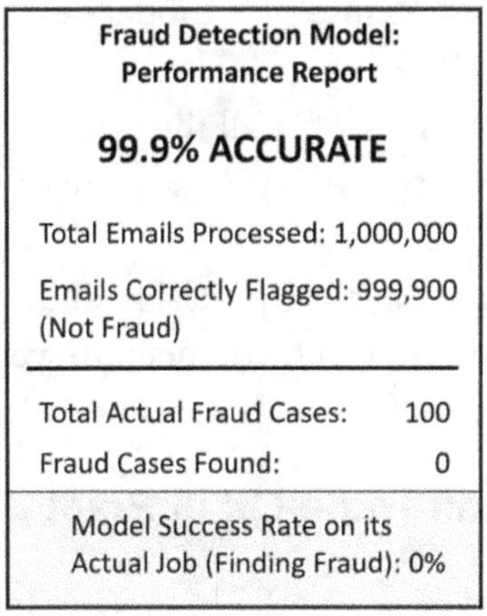

**Fraud Detection Model:
Performance Report**

99.9% ACCURATE

Total Emails Processed: 1,000,000

Emails Correctly Flagged: 999,900
(Not Fraud)

Total Actual Fraud Cases: 100

Fraud Cases Found: 0

Model Success Rate on its
Actual Job (Finding Fraud): 0%

The most intuitive metric is **accuracy**, which measures the overall proportion of correct predictions. However, accuracy can be dangerously misleading, especially in the presence of imbalanced datasets—a phenomenon known as the **accuracy paradox** (Valverde-Albacete & Peláez-Moreno, 2014). This paradox arises when one class is far more frequent than others. Consider a model designed to detect a rare disease that occurs in only 1 in 1,000 patients. A trivial model that simply predicts "no disease" for every single patient will achieve 99.9% accuracy. While technically correct on 999 out of 1,000 cases, this model is completely useless because it fails to identify the one case it was built to find. This high accuracy score gives a false sense of performance and masks the model's total failure on the minority class.

This scenario is extremely common in real-world applications like fraud detection, network intrusion detection, and medical diagnostics. Because of the accuracy paradox, relying on accuracy alone for imbalanced datasets is a critical error. It is for this reason that practitioners must turn to more nuanced metrics like precision and recall, which evaluate a model's performance on the positive class independently of the negative class, providing a much clearer picture of its true utility (Powers, 2011).

15.3.2 Precision, Recall, and the F1-Score

To get a true picture of a model's performance, especially on imbalanced datasets, we must use precision and recall, metrics that have their roots in the field of information retrieval (van Rijsbergen, 1979; Christen et al., 2024; Banerjeee et al., 2024).

- **Precision:** Asks, "When the model predicted the positive class, how often was it correct?" It is the measure of a model's exactness.
  Formula: TP / (TP + FP)
- **Recall (or Sensitivity):** Asks, "Of all the actual positive cases, how many did the model correctly identify?" It is the measure of a model's completeness.
  Formula: TP / (TP + FN)
- **F1-Score:** The harmonic mean of precision and recall. It provides a single, balanced score that is useful when you need to weigh the concerns of both.
  Formula: 2 * (Precision * Recall) / (Precision + Recall)

15.3.3 The Precision-Recall Trade-Off in Practice

In nearly every real-world application, there is an inherent trade-off between precision and recall. Tuning a model to increase its precision (reducing false positives) will often decrease its recall (increasing false negatives), and vice versa (Davis & Goadrich, 2006). The decision of which metric to prioritize is not a technical one, but a business one, and it depends entirely on the cost of each type of error.

Consider a medical diagnostic scenario. For a preliminary cancer screening test, the primary goal is to avoid missing any potential cases. The cost of a **false negative** (misdiagnosing a sick patient as healthy) is catastrophic. The cost of a **false positive** (telling a healthy patient they might be sick) is far lower, as it simply leads to more detailed, follow-up testing. Therefore, for this application, the model would be tuned to maximize **recall**, even at the expense of lower precision. We would rather have more false alarms than miss a single case (Powers, 2011).

Figure 15.2

The Precision-Recall Trade-Off: You must choose which error is less disastrous.

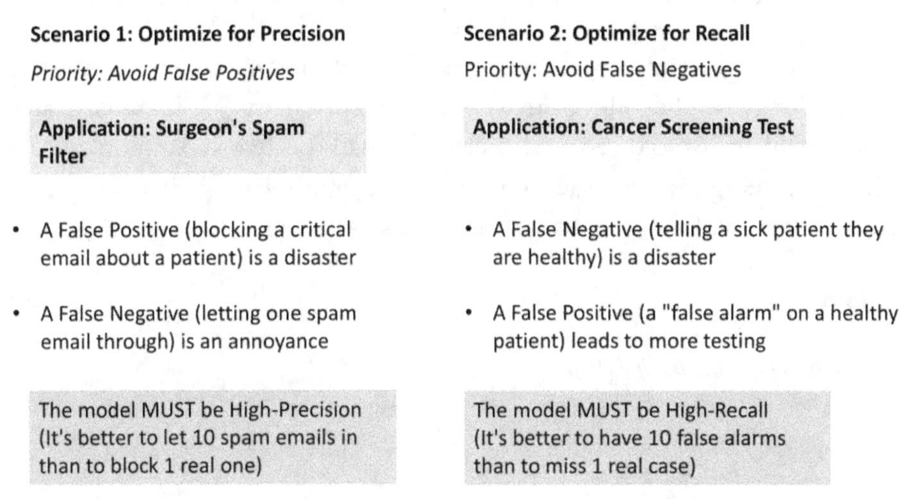

Now, consider the opposite scenario: a spam filter for a surgeon's email inbox. The cost of a **false positive** (incorrectly marking a critical, time-sensitive email from another doctor about a patient's surgery as spam) could be disastrous. The cost of a **false negative** (letting a single piece of junk mail through to the inbox) is merely a minor annoyance. In this case, the model must be tuned to maximize **precision**. We would rather let some spam through than risk losing a single important email. Understanding this trade-off and aligning the model's evaluation metric with the specific business or user needs is one of the most critical skills for a data scientist (Sokolova & Lapalme, 2009).

15.3.4 Advanced Classification Analysis: ROC and AUC

While precision and recall provide evaluation at a single operating point (threshold), the **receiver operating characteristic (ROC)** curve offers a vital, comprehensive view of a classifier's performance across all possible decision thresholds. The ROC curve is a plot of the **true positive rate** (TPR, identical to Recall) against the **false**

positive rate (FPR). The FPR, calculated as

$$\frac{\text{False Positives}}{\text{False Positives} + \text{True Negatives}}$$

represents the proportion of negative cases incorrectly classified as positive.

The ROC curve is particularly useful because it illustrates the inherent trade-off between sensitivity (finding all positive cases) and specificity (avoiding false alarms) across all possible threshold settings. A classifier with perfect separation will have a curve that runs along the upper-left corner of the graph. In contrast, a useless random classifier will follow the diagonal line from (0, 0) to (1, 1).

Figure 15.3

Receiver Operating Characteristic (ROC) Curve

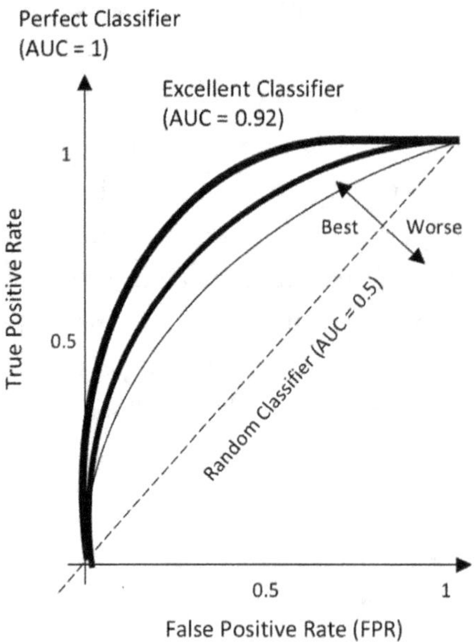

The **area under the curve (AUC)** is the single scalar metric used to summarize the performance of the ROC curve. AUC is the probability that a model ranks a randomly chosen positive instance higher than a randomly chosen negative instance. A perfect classifier achieves an AUC of 1.0, while random guessing yields an AUC of 0.5. Since AUC is independent of the decision threshold, it provides an indispensable measure of a model's intrinsic quality.

15.4 Metrics for Generation and Translation Tasks

Evaluating generated text is more complex, as there is no single correct answer. These tasks rely on metrics that compare the machine-generated text to one or more human-written reference texts. While the field is an active area of research with many specialized metrics, this section focuses on the three most foundational and widely used standards that every NLP practitioner must know: BLEU, ROUGE, and METEOR.

- **BLEU (Bilingual Evaluation Understudy):** The standard for machine translation. It is a precision-focused metric that measures the overlap of N-grams (sequences of words) between the generated text and the reference translations.
    - **Strengths.** It is fast, language-independent, and among the first metrics to show reasonable correlation with human judgment **at the corpus-level** (i.e., when averaging scores over an entire test set). This made it a standard for rapid, iterative development (Papineni et al., 2002).
    - **Weaknesses.** BLEU correlates poorly with human judgments of fluency and adequacy **at the sentence level**, as it does not capture meaning, synonymy, or grammatical correctness. A translation can achieve a high BLEU score even when the words are scrambled but correct (Callison-Burch et al., 2006).
- **ROUGE (Recall-Oriented Understudy for Gisting Evaluation):** The standard for text summarization. As its name suggests, ROUGE is a recall-oriented metric. It measures how many N-grams from the human-created reference summaries are found in the machine-generated summary.
    - **Rationale.** For summarization, it is more important that the system captures all the key points from the reference, making recall a more suitable measure than precision (Lin, 2004).
- **METEOR (Metric for Evaluation of Translation with Explicit Ordering):** An alternative to BLEU that addresses some of its shortcomings. METEOR aligns the candidate and reference translations and

considers synonymy (using resources like WordNet) and stemmed words. It generally correlates better with human judgments of translation quality than BLEU (Banerjee & Lavie, 2005).

15.5 Metrics for Information Retrieval

Information retrieval systems, like search engines, are evaluated on their ability to return a ranked list of relevant documents. This is different from a simple classification because the *order* of the results is critically important.

- **Mean Average Precision (MAP)**: A widely-used metric that evaluates the overall quality of a ranked list. MAP's main assumption is binary relevance: a document is either relevant (1) or not (0).
    - To calculate MAP, we first find the average precision (AP) for each individual query. AP is the average of the precision scores calculated *at the rank of each relevant document.*
    - For example, if a query returns five results and the relevant documents are at ranks 1, 3, and 5, we would calculate the precision at each of those ranks:
        - Rank 1: Precision = 1/1 = 1.0
        - Rank 3: Precision = 2/3 = 0.67
        - Rank 5: Precision = 3/5 = 0.6
    - The average precision (AP) for *this query* would be (1.0 + 0.67 + 0.6) / 3 = 0.76.
    - The mean average precision (MAP) is then the mean of these AP scores across *all* queries in the test set (Manning et al., 2008).
- **Normalized Discounted Cumulative Gain (NDCG)**: A more sophisticated ranking metric that improves on MAP's primary weakness. NDCG assumes graded relevance, meaning a document can be more than just 'relevant' or 'not'—it can be 'okay' (1), 'good' (2), or 'perfect' (3).
    - It operates in two stages:
        1. **Discounted Cumulative Gain (DCG)**: It rewards placing highly relevant documents at the top of the list. A 'perfect' (3) document at rank 1 contributes more to the score than a 'perfect' (3) document at rank 10, which is 'discounted'.
        2. **Normalized (NDCG)**: The DCG score is then normalized (divided) by the 'ideal' DCG score (the score of a perfect ranking). This ensures the final score is always between 0.0

and 1.0.

- ○ NDCG is the preferred metric in modern search engines because it aligns better with user satisfaction, correctly capturing the idea that finding a 'perfect' document is much better than finding a 'good' one, and finding either at the top of the list is best (Järvelin & Kekäläinen, 2002).

Figure 15.4

How NDCG (Ranking Metric) Works: Why the Order of Search Results Matters

Search Result A (Low NDCG Score)

Rank 1 : [Bad Document]
Relevance = 0

Rank 2 : [Good Document]
Relevance = 1

Rank 3 : [Perfect Document]
Relevance = 2

Problem: The most valuable result ("Perfect Document") is at the bottom. Its value is "discounted" by its low rank.

Search Result B (High NDCG Score)

Rank 1 : [Perfect Document]
Relevance = 2

Rank 2 : [Good Document]
Relevance = 1

Rank 3 : [Bad Document]
Relevance = 0

Advantage: The most valuable result is at Rank 1, where it gets full value (no discount). This result is scored much higher.

15.6 The Gold Standard: Human Evaluation

Despite their utility, automatic metrics remain imperfect proxies for true quality. The ultimate judge of an NLP system's performance is a human. Human evaluation is essential for assessing qualities that automatic metrics cannot capture, such as fluency, adequacy, coherence, and helpfulness (Callison-Burch et al., 2007).

However, conducting a reliable human evaluation is itself a rigorous scientific process. It is not enough to simply ask a few people for their opinion. A robust process involves several key steps:

- **Annotation Guidelines:** A detailed set must be created. This document is the "source of truth" for the evaluators, defining precisely what constitutes a 'good' or 'bad' output and providing clear examples for edge cases.
- **Inter-Annotator Agreement (IAA):** Multiple independent raters (typically

at least three) are used to label the same set of outputs. Their ratings are then compared to measure the inter-annotator agreement (IAA), using statistical measures like Cohen's kappa or Fleiss' kappa (Artstein & Poesio, 2008). A high IAA indicates that the guidelines are clear and the task is well-defined, which means the resulting data are reliable.

Only after establishing reliable human judgments can we confidently assess the true quality of an NLP system.

15.6.1 Technical Note: The Evolving Role of Crowdsourcing

The entire process of gathering and aggregating reliable human scores is often managed on crowdsourcing platforms—online marketplaces that distribute human intelligence tasks (HITs) to a large, decentralized workforce. Amazon Mechanical Turk (AMT) is the historical and foundational example of this marketplace model (Snow et al., 2008).

In the age of large language models (LLMs), the function of crowdsourcing has shifted toward assessing complex judgments:

- **Establishing the Gold Standard:** Crowdsourcing remains indispensable for setting the human benchmark for metrics such as fluency and helpfulness against which complex AI outputs are measured.
- **Focus on Complex Evaluation:** Platforms are crucial for assessing nuances too subtle for automation, such as comparing two LLM outputs side-by-side or auditing text for factual hallucination and algorithmic bias.
- **Industry Alternatives:** To meet higher quality and security standards (common in financial or medical domains), researchers often rely on specialized annotation services or internal, managed workforces that provide dedicated quality control and domain expertise, moving beyond the general marketplace model.

15.7 Conclusion

Evaluation is the compass that guides progress in NLP. It is the rigorous discipline that allows us to move from "it seems to work" to "we can prove it is better." This chapter has explored the two primary philosophies of evaluation—intrinsic and extrinsic—and detailed the specific metrics used for the field's most common tasks. We have seen why simple accuracy can be misleading and why a deep understanding of the precision-recall trade-off is essential for any real-world application. We have also established that while automatic metrics like BLEU and ROUGE are invaluable for rapid experimentation, the ultimate measure of quality remains human judgment, which must itself be collected through a scientifically rigorous process. A successful NLP project requires a thoughtful combination of these approaches: rapid, iterative development using intrinsic, automatic metrics, followed by careful, application-oriented, and rigorous, extrinsic evaluation with human-in-the-loop assessment before deployment. A deep understanding of this evaluation toolkit is what separates a good practitioner from a great one, enabling them not only to build models but also to prove their worth.

15.8 References

Artstein, R., & Poesio, M. (2008). Inter-coder agreement for computational linguistics. *Computational Linguistics, 34*(4), 555-596. https://doi.org/10.1162/coli.07-034-R2

Banerjee, S., & Lavie, A. (2005). METEOR: An automatic metric for MT evaluation with improved correlation with human judgments. *In Proceedings of the ACL Workshop on Intrinsic and Extrinsic Evaluation Measures for Machine Translation and/or Summarization* (pp. 65–72).

Banerjee, D., Sharma, N., Chauhan, R., Singh, M., & Kumar, B. V. (2024). Improving precision in rose leaf disease recognition with integrated CNN and SVM models. *2024 4th International Conference on Innovative Practices in Technology and Management* (ICIPTM), 1–6. https://doi.org/10.1109/ICIPTM59628.2024.10563533

Callison-Burch, C., Fordyce, C., Koehn, P., Monz, C., & Schroeder, J. (2007). (Meta-) evaluation of machine translation. *In Proceedings of the Second Workshop on Statistical Machine Translation* (pp. 136–158).

Callison-Burch, C., Osborne, M., & Koehn, P. (2006). Re-evaluating the role of BLEU in machine translation research. *In Proceedings of the 11th Conference of the European Chapter of the Association for Computational Linguistics (EACL 2006)* (pp. 249–256).

Christen, P., Hand, D. J., & Kirielle, N. (2024). A review of the F-measure: Its history, properties, criticism, and alternatives. *ACM Computing Surveys, 56*(3), Article 73. https://doi.org/10.1145/3606367

Davis, J., & Goadrich, M. (2006). The relationship between precision-recall and ROC curves. *In Proceedings of the 23rd International Conference on Machine Learning* (pp. 233–240). https://doi.org/10.1145/1143844.1143874

Fawcett, T. (2006). An introduction to ROC analysis. *Pattern Recognition Letters, 27*(8), 861–874. https://doi.org/10.1016/j.patrec.2005.10.010

Järvelin, K., & Kekäläinen, J. (2002). Cumulated gain-based evaluation of IR techniques. *ACM Transactions on Information Systems, 20*(4), 422–446. https://doi.org/10.1145/582415.582418

Jurafsky, D., & Martin, J. H. (2023). *Speech and language processing* (3rd ed.). Prentice Hall.

Lin, C. Y. (2004). ROUGE: A package for automatic evaluation of summaries. In Text summarization branches out: *Proceedings of the ACL-04 workshop* (pp. 74–81). Association for Computational Linguistics. https://aclanthology.org/W04-1013/

Manning, C. D., & Schütze, H. (1999). *Foundations of statistical natural language processing.* MIT Press.

Manning, C. D., Raghavan, P., & Schütze, H. (2008). *Introduction to information retrieval.* Cambridge University Press. https://doi.org/10.1017/CBO9780511809071

Papineni, K., Roukos, S., Ward, T., & Zhu, W. J. (2002). BLEU: A method for automatic evaluation of machine translation. *In Proceedings of the 40th Annual Meeting of the Association for Computational Linguistics* (pp. 311–318). https://doi.org/10.3115/1073083.1073135

Powers, D. M. (2011). Evaluation: from precision, recall and F-measure to ROC, informedness, markedness & correlation. *Journal of Machine Learning Technologies, 2*(1), 37–63. https://arxiv.org/abs/2010.16061

Snow, R., O'Connor, B., Jurafsky, D., & Ng, A. Y. (2008). Cheap and fast: How crowdsourcing can enable more NLP research. *In Proceedings of the 46th Annual Meeting of the Association for Computational Linguistics on Human Language Technologies: Short Papers* (pp. 25–28).

Sparck Jones, K., & Galliers, J. R. (1996). Evaluating natural language processing systems: An analysis and review. Springer.

Valverde-Albacete, F. J., & Peláez-Moreno, C. (2014). 100% classification accuracy is not enough: The accuracy paradox in the evaluation of sensor-based diagnostic systems. *In Proceedings of the 10th International Conference on Intelligent Data Acquisition and Advanced Computing Systems: Technology and Applications.*

van Rijsbergen, C. J. (1979). *Information retrieval* (2nd ed.). Butterworth-Heinemann.

15.9 Glossary

Accuracy (Definition): A classification metric measuring the proportion of all predictions that were correct.

Adequacy/Fidelity: A criterion in human evaluation, typically for translation or summarization, that assesses how well the generated text preserves the meaning of the original source text.

AUC (Area Under the Curve): A metric that summarizes a classifier's performance across all classification thresholds, derived from the ROC curve.

BLEU (Bilingual Evaluation Understudy): A precision-focused metric used to evaluate the quality of machine-translated text by comparing N-gram overlap with reference translations.

Coherence: A criterion in human evaluation that assesses how well the sentences in a generated text connect to each other in a logical and understandable way.

Cohen's Kappa: A statistic used to measure inter-annotator agreement for categorical items between two raters, accounting for the possibility of agreement occurring by chance.

Confusion Matrix: A table used to describe the performance of a classification model, showing the counts of true positives, true negatives, false positives, and false negatives.

Extrinsic Evaluation: An evaluation of an NLP component based on its impact on the performance of a larger, downstream application.

F1-Score: The harmonic mean of precision and recall, providing a single, balanced score for a classification model.

Fleiss' Kappa: A statistical measure for assessing the reliability of agreement between three or more raters when making categorical judgments. It is a generalization of cohen's kappa for multiple raters.

Fluency: A criterion in human evaluation that assesses how grammatically correct, natural, and easy to read a piece of machine-generated text is.

Ground-Truth Dataset: A dataset used for evaluation that contains inputs paired with their corresponding correct, human-verified outputs (the "ground truth"). This dataset is treated as the "correct answer key" against which the model's performance is measured. It is also often referred to as a "gold standard."

Helpfulness: A criterion in human evaluation, often for dialogue systems or chatbots, that assesses whether the model's response successfully addresses the user's goal or solves their problem.

Inter-Annotator Agreement (IAA): A measure of how much consensus there is between two or more human raters when labeling data.

Intrinsic Evaluation: An evaluation of an NLP model on a specific, isolated subtask against a ground-truth dataset.

MAP (Mean Average Precision): A metric for evaluating ranked information retrieval results. It is the mean of the "average precision" (AP) scores across all queries. AP is calculated for a single query by averaging the precision scores at the rank of each relevant document. It assumes binary relevance (a document is either relevant or not).

METEOR (Metric for Evaluation of Translation with Explicit Ordering): An alternative to BLEU that performs an alignment and considers synonymy (using resources like WordNet) and stemming, often correlating better with human judgment at the sentence-level.

NDCG (Normalized Discounted Cumulative Gain): A measure of ranking quality that improves on MAP by using graded relevance (e.g., 0=bad, 1=okay, 2=good). It rewards placing highly relevant documents at the top of the list by applying a logarithmic 'discount' to the value of documents as they appear lower in the ranking.

Precision: A classification metric measuring a model's exactness. It answers the question: "Of all the times the model predicted 'positive,' how often was it actually correct?"

Recall (Sensitivity): A classification metric measuring a model's completeness. It answers the question: "Of all the actual positive cases that exist, how many did the model successfully find?"

ROC (Receiver Operating Characteristic) Curve: A graph showing the performance of a binary classifier at all classification thresholds, plotting the true positive rate against the false positive rate.

ROUGE (Recall-Oriented Understudy for Gisting Evaluation): A recall-focused set of metrics used to evaluate automatic summarization by comparing N-gram overlap with reference summaries.

15.10 Review Questions and Discussion Topics

1. **Metric Selection:** You are developing an NLP model to scan medical reports and identify mentions of adverse drug reactions. The goal is to ensure no potential reaction is missed. Which metric—precision or recall—is more critical for this task, and why? Describe the real-world consequences of an error related to each metric.

2. **BLEU Score Limitations:** A machine translation system translates the English sentence "The cat sat on the mat" into the French "le le le le le le" (the word 'the' repeated six times). A second system translates it to a valid, synonymous French sentence like "Un félin reposait sur le tapis" ("A feline was resting upon the rug"). Which translation would likely get a higher BLEU score (assuming a standard reference like "Le chat s'assit sur le tapis"), and why? What does this tell you about the limitations of BLEU?

3. **Intrinsic Vs. Extrinsic:** You have developed a new algorithm for part-of-speech tagging. Describe how you would perform both an intrinsic and an extrinsic evaluation of your new model. What are the pros and cons of each approach?

4. **Human Evaluation Design:** You need to evaluate a new chatbot designed for customer service. What key qualities (e.g., fluency, helpfulness) would you ask human evaluators to score? How would you design a rating scale (e.g., 1-5) for each quality? Why is calculating inter-annotator agreement important for this process?

5. **The Problem with Accuracy:** A bank wants to build a model to detect fraudulent transactions. Fraudulent transactions account for only 0.1% of all transactions. A junior data scientist builds a model and reports that it has 99.9% accuracy. Why is this metric dangerously misleading? Which metrics would provide a more truthful picture of the model's performance?

Chapter 16

Information Extraction: From Raw Text to Actionable Knowledge

16.1 Introduction: From Text Analysis to Information Extraction

As established in the previous chapters, the ability to derive meaning from unstructured text relies on a suite of core analytical methods. Among the most powerful of these is information extraction (IE), a cornerstone of modern natural language processing (NLP) that automates the transformation of unstructured text into structured, machine-readable formats. This process is fundamental for enabling artificial intelligence systems to comprehend, process, and reason over the vast quantities of information embedded in documents, reports, and web pages.

The primary objective of IE is to automatically identify and classify user-defined entities, relationships, and events. What constitutes 'important' information depends heavily on the specific application, ranging from tracking financial transactions to monitoring disease outbreaks. Unlike text classification or sentiment analysis, which assign a single label to an entire document, IE operates at a much more granular level. It dissects individual sentences to pinpoint specific pieces of data and, crucially, the relationships that connect them. This process transforms a sea of words into a structured network of facts—a knowledge base or knowledge graph—that can fuel advanced analytics and logical reasoning.

This chapter provides a detailed examination of the IE pipeline, focusing on the foundational tasks of named entity recognition (NER), coreference resolution, and relation detection and classification.

16.2 Named Entity Recognition (NER): The Foundation

Named entity recognition (NER) is the foundational task in IE, focusing on identifying and classifying specific entities—or rigid designators—in text into predefined categories. These categories typically include persons, organizations, locations, dates, and monetary values (Nadeau & Sekine, 2007; Reddy et al., 2023). In essence, NER pinpoints the 'who,' 'what,' 'when,' and 'where' in a document. The accuracy of the entire IE pipeline is heavily dependent on the performance of this initial step.

16.2.1 Challenges in NER: Ambiguity and Variation

A significant challenge in NER is the inherent ambiguity of natural language. Words can have multiple meanings (polysemy) or share spellings with other words (homonymy). For example, 'Washington' could refer to a person, a state, or a city. 'Ford' could be a person's name or a car company. A successful NER system must effectively use the surrounding context to disambiguate these cases (Wang & Iwaihara, 2019). Furthermore, entities can be expressed in countless variations. "International Business Machines," 'IBM,' and "Big Blue" all refer to the same organization. An effective NER system must be robust enough to recognize these different surface forms as belonging to the same entity.

16.2.2 NER as a Sequence Labeling Task

The dominant approach to NER is to frame it as a sequence labeling task. The goal is to assign a label to each token in a sentence that indicates whether it is part of a named entity and, if so, what type. The most common tagging scheme is the **IOB (Inside, Outside, Beginning)** format (Ramshaw & Marcus, 1995; Liu et al., 2022):

- **I-type:** Marks a token that is inside an entity

- **O:** Marks a token that is outside any entity

- **B-type:** Marks the beginning of a named entity

For example, "Arvind Krishna works for IBM" would be tagged:

Arvind (B-PERSON) Krishna (I-PERSON) works (O) for (O) IBM (B-ORGANIZATION)

Figure 16.1

NER as a Sequence Labeling Task: Using the IOB Tagging Scheme

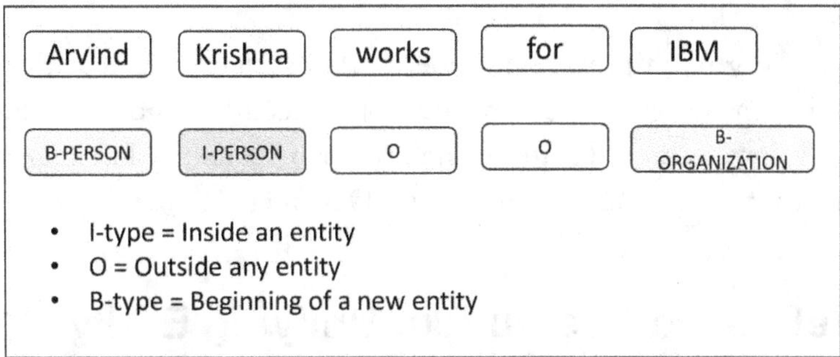

Framing NER in this way is powerful because it allows the model to learn not just the properties of individual words, but also the dependencies between adjacent tags. This is where modern neural architectures excel. A common and highly effective model for this task is a Bidirectional LSTM combined with a conditional random field layer (BiLSTM-CRF) (Huang et al., 2015). The **BiLSTM** part of the model is crucial for learning context. By processing the sentence in both forward and backward directions, the representation for each word contains information about the entire sentence. This helps the model disambiguate; for example, seeing the word 'Inc.' later in a sentence can help it correctly identify an earlier word as part of an organization's name. The **CRF** layer acts as a final "grammar checker" on top of the BiLSTM's predictions. During training, the CRF layer learns the constraints of the tagging scheme. For instance, it learns that an I-PERSON tag is highly unlikely to follow a B-ORGANIZATION tag, or that an I-PERSON tag cannot appear after an O tag. By considering the entire sequence of predicted tags together, the CRF ensures that the final output is not just a series of locally plausible tags, but a globally coherent and valid sequence (Lafferty et al., 2001).

16.2.3 NER Architectures: From Rules to Transformers

- **Rule-based Systems:** These early systems use handcrafted rules, regular expressions, and dictionaries (gazetteers) to identify entities. They are precise in narrow domains but are brittle and labor-intensive to maintain (Grishman, 2019).

- **Classic Machine Learning:** Models like hidden Markov models (HMMs) and conditional random fields (CRFs) represented a major advance. CRFs, in particular, became a standard due to their ability to incorporate a rich set of engineered features (e.g., capitalization, part-of-speech tags) to make globally optimal tagging decisions (Mohri et al., 2018).

- **Deep Learning Architectures:** Modern NER systems, particularly those based on the BiLSTM-CRF architecture and, more recently, large pre-trained models like BERT, have achieved state-of-the-art performance by learning feature representations automatically from data (Devlin et al., 2019).

16.3 Coreference Resolution: Unifying Entity Mentions

Once NER has identified the initial entity mentions, the next step is to figure out which of these mentions refer to the same real-world entity. This is the task of **coreference resolution** (Jurafsky & Martin, 2023). It is a notoriously difficult problem in NLP because of the sheer variety of ways we refer to things. Consider the text: "**Arvind Krishna** took over as CEO of **IBM. He** succeeded Ginni Rometty. **The new chief executive** is leading **the company** through a major transformation." Coreference resolution would need to determine that "**Arvind Krishna**," 'He,' and "**The new chief executive**" all refer to the same person, and that "**IBM**" and "**the company**" refer to the same organization.

The challenges are immense. Resolving pronouns like 'he' or 'it' requires understanding grammatical constraints (e.g., gender agreement) and contextual clues. Resolving definite noun phrases like "the new chief executive" requires both discourse context (knowing who was just mentioned) and world knowledge (knowing that a CEO is a type of executive) (Hobbs, 1978; Prolo, 2006). The distance between mentions can also be very large, spanning multiple paragraphs or even pages. Early systems relied on complex, hand-crafted rules and heuristics (Soon et al., 2001; Ren et al., 2008). Modern systems, however, have increasingly turned to deep learning, training models to learn a similarity score between pairs of potential entity mentions. These models take into account a rich set of features, including the text between the mentions, their grammatical roles, and their vector representations, to make more accurate and robust coreference decisions (Lee et al., 2017).

Figure 16.2

Coreference Resolution

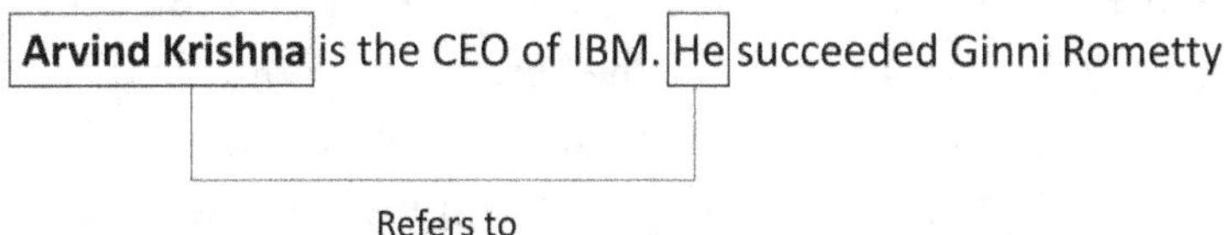

16.4 Relation Detection and Classification: Connecting the Dots

After identifying and unifying entities, the next crucial step is **relation detection and classification**. This task identifies and categorizes the semantic relationships that exist between these entities. It transforms a simple list of entities into a structured network of interconnected knowledge (Jurafsky & Martin, 2023). For example, given the NER output from "Bill works for IBM," this step would identify the works-for(Bill, IBM) relationship. The types of relations are predefined based on the application's needs, such as located-in(Person, Location) or subsidiary-of (Organization, Organization).

There are several dominant paradigms for this task. **Pattern-based or rule-based systems** rely on handwritten linguistic patterns (e.g., [PERSON] works for [ORGANIZATION]) to identify relations. These are highly precise but suffer from low recall, as they cannot capture the countless ways a single relationship can be expressed in language (Hearst, 1992; Kawamura et al., 2017). **Supervised machine learning approaches** treat the problem as a classification task. A model is trained on a large corpus of sentences that have been manually annotated with entities and the relations between them. While these models can achieve high accuracy, the need for large, hand-labeled datasets is a major bottleneck (Zelenko et al., 2003; Choi et al, 2014).

To overcome this, **distant supervision** has become a popular and scalable alternative. This semi-supervised method automatically generates a large, albeit *noisy*, training dataset. The process is a clever, high-speed shortcut. It begins by taking a large knowledge base of known facts, such as (Steve Jobs, founded Apple), from a source like Wikidata. It then scans a massive, unannotated text corpus (like a news archive) and applies a simple, powerful assumption. It assumes that any sentence containing *both* "Steve Jobs" and 'Apple' is a **"positive training example"** for the founded relation (Ji et al., 2017).

A **"positive training example"** is simply a 'yes' example—a specimen of *exactly* what the model is supposed to learn to find. This assumption is 'noisy' because it's often wrong. For instance, the system might correctly label the sentence "Steve Jobs founded Apple in 1976." This is a true positive example, and it teaches the model a valid pattern. However, the system would *incorrectly* label the sentence "A new documentary about Steve Jobs was screened at the Apple campus" as a positive example, even though that sentence has nothing to do with the founded relation. The central bet of distant supervision is that the correct, useful examples will be statistically dominant, allowing the model to learn the true patterns *despite* the noise from all the incorrect ones (Mintz et al., 2009; Smirnova & Cudré-Mauroux, 2019).

16.5 Event and Temporal Extraction: Capturing Dynamic Information

Beyond static relationships, IE systems are often tasked with extracting information about events and when they occurred. **Event extraction** identifies occurrences of specific types of events in a text and extracts their participants (arguments) and roles. For example, in a sentence about a corporate acquisition, the system would identify the 'acquisition' event and extract the acquirer, the acquired, and the price (Leeuwenberg & Moens, 2019). **Temporal extraction** then anchors these events in time by identifying and normalizing temporal expressions like 'yesterday,' "October 2023," or "next quarter" (Pustejovsky et al., 2003; Pustejovsky, 2017).

16.6 Practical Application: Building a Knowledge Graph with IE

The ultimate goal of a comprehensive IE pipeline is often the creation of a **knowledge graph (KG)**. A KG is a structured representation of knowledge where nodes represent entities and labeled, directed edges represent the relationships between them (Hogan et al., 2021). The process integrates all the preceding steps: NER identifies the nodes, coreference resolution merges them, and Relation/Event Extraction identifies the labeled edges. The resulting graph provides a structured, comprehensive representation of the knowledge contained in the source documents.

The value of a KG is immense. It transforms a collection of unstructured documents into a queryable knowledge base. For example, a KG built from financial news could power sophisticated search queries like, "Find all CEOs who used to work for IBM and are now at companies that were acquired by Microsoft." This type of multi-hop reasoning is impossible with standard text search but is a natural operation on a graph structure. Knowledge graphs are the backbone of many advanced AI applications, including Google's Knowledge Panel, which provides structured answers to search queries, as well as sophisticated recommendation engines and systems for drug discovery that can reason over complex networks of proteins, diseases, and chemical compounds (Ji et al., 2022). They represent the culmination of the IE process: the successful transformation of raw, unstructured text into actionable, interconnected knowledge.

Figure 16.3

From Text to Knowledge Graph

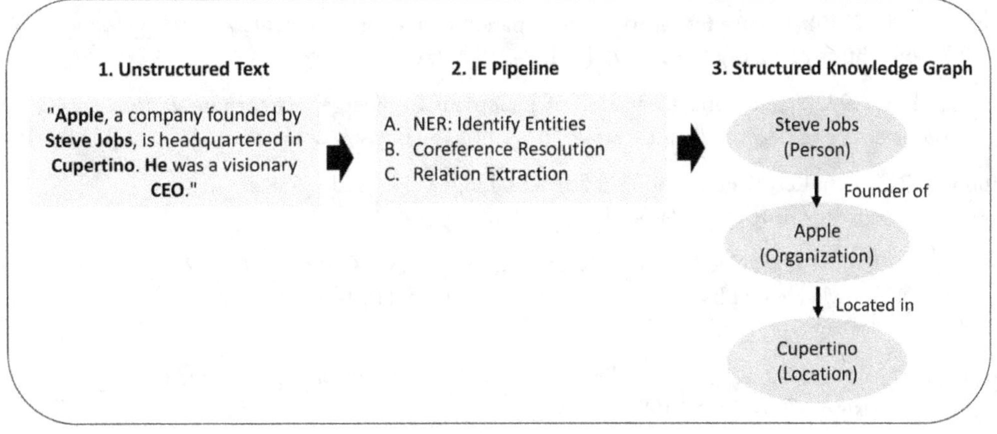

16.7 Conclusion: From Unstructured Text to Actionable Knowledge

Information extraction provides the critical bridge between the world's vast repository of unstructured text and the structured data needs of intelligent systems. The process is a sequential pipeline, beginning with the foundational task of named entity recognition to identify key entities, a step that must overcome significant challenges of linguistic ambiguity. Following NER, coreference resolution unifies entity mentions, while relation and event extraction build a network of connections, transforming a simple list of entities into a meaningful graph of knowledge. The evolution of IE architectures, from brittle rule-based systems to robust, deep learning models, has dramatically increased the accuracy and scalability of this process. The ultimate output, a knowledge graph, enables the kind of complex, multi-step reasoning that is a hallmark of true artificial intelligence, powering the next generation of search, analytics, and discovery applications (Hogan et al., 2021).

16.8 References

Choi, S.-P., Lee, S., Jung, H., & Song, S. (2014). An intensive case study on kernel-based relation extraction. *Multimedia Tools and Applications*, 71(2), 741–767. https://doi.org/10.1007/s11042-013-1380-5

Devlin, J., Chang, M. W., Lee, K., & Toutanova, K. (2019). BERT: Pre-training of deep bidirectional transformers for language understanding. *In Proceedings of the 2019 Conference of the North American Chapter of the Association for Computational Linguistics: Human Language Technologies*, Volume 1 (Long and Short Papers) (pp. 4171-4186). https://doi.org/10.18653/v1/N19-1423

Grishman, R. (2019). Twenty-five years of information extraction. *Natural Language Engineering, 25*(6), 677–690. https://doi.org/10.1017/S1351324919000512

Hearst, M. A. (1992). Automatic acquisition of hyponyms from large text corpora. *In Proceedings of the 14th Conference on Computational Linguistics* (pp. 539-545). https://doi.org/10.3115/992133.992154

Hobbs, J. R. (1978). Resolving pronoun references. Lingua, 44(4), 311–338. https://doi.org/10.1016/0024-3841(78)90006-2

Hogan, A., Blomqvist, E., Cochez, M., d'Amato, C., de Melo, G., Gutierrez, C., ... & Zimmermann, A. (2021). Knowledge graphs. ACM Computing Surveys, 54(4), 1–37. https://doi.org/10.1145/3447772

Huang, Z., Xu, W., & Yu, K. (2015). Bidirectional LSTM-CRF models for sequence tagging. https://arxiv.org/abs/1508.01991

Ji, G., Liu, K., He, S., & Zhao, J. (2017). Distant supervision for relation extraction with sentence-level attention and entity descriptions [C]. *Proceedings of the ... AAAI Conference on Artificial Intelligence, 31*(1). https://doi.org/10.1609/aaai.v31i1.10953

Ji, S., Pan, S., Cambria, E., Marttinen, P., & Yu, P. S. (2022). A survey on knowledge graphs: Representation, acquisition, and applications. *IEEE Transactions on Neural Networks and Learning Systems, 33*(2), 494–514. https://doi.org/10.1109/TNNLS.2021.3070843

Jurafsky, D., & Martin, J. H. (2023). Speech and language processing (3rd ed.). Prentice Hall.

Kawamura, T., Sekine, M., & Matsumura, K. (2017). Detecting hypernym/hyponym in science and technology thesaurus using entropy-based clustering of word vectors. *International Journal of Semantic Computing, 11*(4), 433–449. https://doi.org/10.1142/s1793351x17400177

Lafferty, J., McCallum, A., & Pereira, F. C. (2001). Conditional random fields: Probabilistic models for segmenting and labeling sequence data. *Proceedings of the Eighteenth International Conference on Machine Learning*, 282-289.

Lee, K., He, L., & Zettlemoyer, L. (2017). End-to-end neural coreference resolution. *In Proceedings of the 2017 Conference on Empirical Methods in Natural Language Processing* (pp. 188-197). https://doi.org/10.18653/v1/D17-1018

Leeuwenberg, A., & Moens, M.-F. (2019). A Survey on Temporal Reasoning for Temporal Information Extraction from Text. *The Journal of Artificial Intelligence Research*, 66, 341–380. https://doi.org/10.1613/jair.1.11727

Liu, P., Guo, Y., Lei, J., & Li, G. (2022). Tagging schemes can do more in named entity recognition: Take Chinese as an example. *2022 International Conference on Networks, Communications and Information Technology* (CNCIT), 187–191. https://doi.org/10.1109/CNCIT56797.2022.00037

Mintz, M., Bills, S., Snow, R., & Jurafsky, D. (2009). Distant supervision for relation extraction without labeled data. *In Proceedings of the Joint Conference of the 47th Annual Meeting of the ACL and the 4th International Joint Conference on Natural Language Processing of the AFNLP* (pp. 1003-1011).

Mohri, M., Rostamizadeh, A., & Talwalkar, A. (2018). *Foundations of machine learning* (2nd ed.). The MIT Press.

Nadeau, D., & Sekine, S. (2007). A survey of named entity recognition and classification. *Lingvisticae Investigationes*, *30*(1), 3-26.

Prolo, C. A. (2006). Computational approaches to pronoun resolution. *Letras de Hoje*, *41*(2), 139–188.

Pustejovsky, J. (2017). ISO-TimeML and the annotation of temporal information. In J. Pustejovsky & N. Ide (Eds.), *Handbook of Linguistic Annotation* (pp. 941–968). Springer Netherlands. https://doi.org/10.1007/978-94-024-0881-2_35

Pustejovsky, J., Castano, J., Ingria, R., Sauri, R., Gaizauskas, R., Setzer, A., & Katz, G. (2003). TimeML: Robust specification of event and temporal expressions in text. In *New directions in question answering* (pp. 28-34).

Ramshaw, L. A., & Marcus, M. P. (1995). Text chunking using transformation-based learning. https://doi.org/10.48550/arxiv.cmp-lg/9505040

Reddy, B. V., Rao, K. S., & Neeraja, K. (2023). Named entity recognition on different languages: A survey. In M. V. Reddy, M. S. Gupta, & A. V. Anand (Eds.), *AIP Conference Proceedings* (Vol. 2492, Number 1). American Institute of Physics. https://doi.org/10.1063/5.0113210

Ren, F., Zhu, J., Wang, H., & Xiao, T. (2008). An effective approach for coreference resolution. *International Journal of Computer Processing of Languages*, 21(2), 135–149. https://doi.org/10.1142/S1793840608001810

Smirnova, A., & Cudré-Mauroux, P. (2019). Relation extraction using distant supervision: A survey. *ACM Computing Surveys*, *51*(5), Article 106. https://doi.org/10.1145/3241741

Soon, W. M., Ng, H. T., & Lim, D. C. Y. (2001). A machine learning approach to coreference resolution of noun phrases. Computational Linguistics, 27(4), 521–544. https://doi.org/10.1162/089120101753342653

Wang, Q., & Iwaihara, M. (2019). Deep neural architectures for joint named entity recognition and disambiguation. *2019 IEEE International Conference on Big Data and Smart Computing (BigComp)*, 1–4.

https://doi.org/10.1109/bigcomp.2019.8679233

Zelenko, D., Aone, C., & Richardella, A. (2003). Kernel methods for relation extraction. *Journal of Machine Learning Research, 3*, 1083–1106.

16.9 Glossary

BERT (Bidirectional Encoder Representations from Transformers): A highly influential pre-trained language model developed by Google that utilizes a deep, bidirectional transformer encoder. Its key innovation is processing an entire sequence left-to-right and right-to-left simultaneously, generating a rich, contextualized representation for every word. It is widely used for analysis and extraction tasks, including named entity recognition (NER).

BiLSTM-CRF: An effective deep learning architecture for sequence labeling tasks like NER, combining a bidirectional LSTM to learn context-aware features and a CRF layer to ensure the final tag sequence is globally optimal.

Conditional Random Fields (CRF): A classical machine learning model used for sequence labeling that considers the context of the entire sequence to make more globally coherent predictions than models that label tokens independently.

Coreference Resolution: The task of identifying all mentions in a text that refer to the same real-world entity and grouping them.

Distant Supervision: A semi-supervised method for creating large-scale training datasets for relation extraction by assuming that if two entities are related in a knowledge base, any sentence containing both entities is a potential training example.

Event Extraction: The process of identifying occurrences of specific types of events in a text and extracting their participants (arguments) and roles.

Information Extraction (IE): The automated process of extracting structured information (such as entities, relationships, and events) from unstructured or semi-structured text.

Google's Knowledge Panel: The high-visibility information box that appears in Google Search results when a user queries for a recognized entity (person, place, organization). It is the most prominent public application powered by the Google Knowledge Graph, serving as the final, structured output of the Information Extraction (IE) process.

IOB (Inside, Outside, Beginning) format: A common tagging scheme for sequence labeling tasks like NER, used to mark whether a token is inside (I), outside (O), or at the beginning (B) of a named entity.

Knowledge Graph (KG): A structured representation of knowledge as a graph, where nodes represent entities and edges represent the relationships between them.

Named Entity Recognition (NER): A subtask of IE that seeks to locate and classify named entities in text into predefined categories such as persons, organizations, locations, etc.

Relation Detection and Classification: The task of identifying and categorizing the semantic relationships that exist between pairs of named entities in a text.

Sequence Labeling: A type of pattern recognition task where the goal is to assign a categorical label to each member of a sequence of observed values (e.g., assigning an IOB tag to each word in a sentence).

Temporal Extraction: The task of identifying and normalizing time expressions in a text to a standard format.

Wikidata: A free, collaborative, multilingual, and structured knowledge base that serves as centralized storage for structured data from the Wikimedia projects. Unlike Wikipedia (which contains unstructured text), Wikidata stores facts in a machine-readable format (e.g., triples of Entity, Property, Value). It is an essential external resource for distant supervision and building large-scale knowledge graphs (KGs).

16.10 Review Questions and Discussion Topics

.1. **Error Propagation:** The IE pipeline is sequential. Discuss the cascading effect of an error in the initial NER stage. If an NER system incorrectly tags "Apple Inc." as just 'Apple,' how does this error impact relation extraction?

2. **Domain and Language Specificity:** An NER system trained on news articles will likely perform poorly on biomedical research papers. Discuss the challenges of domain adaptation for IE systems and the unique challenges of developing IE for low-resource languages.

3. **The Future of IE with Large Language Models (LLMs):** LLMs can perform 'zero-shot' IE tasks via prompting (e.g., "Extract all person and organization names..."). Does this paradigm make traditional, supervised IE pipelines obsolete? Discuss the pros and cons of using LLMs for IE versus fine-tuning smaller, specialized models.

Chapter 17

Applying Natural Language Processing for Social Business Intelligence

17.1 Introduction

The previous chapters provided a detailed examination of the core techniques for text analysis and information extraction, establishing the process for identifying who did what, when, and where, and how to connect those facts into a coherent knowledge graph. This chapter builds directly upon that foundation by applying these techniques to one of the most dynamic, noisy, and commercially valuable data sources available: social media. The practice of analyzing this data, known as social business intelligence (SBI), moves beyond simple fact extraction. It requires us to not only identify entities and relationships but also to understand sentiment, emotion, and intent. This chapter will explore the unique challenges of social media text and the complete suite of NLP tools required to derive true business value.

17.2 The Unique Challenge of Social Media Text

Unlike the well-edited text found in news articles or financial reports, the language of social media is inherently noisy and presents unique challenges for NLP models. Effective analysis requires a robust preprocessing pipeline (as discussed in Chapter 4) to handle the "short, noisy, context-dependent, and dynamic" nature of user-generated content (Li et al., 2022). These challenges include:

- **Informal Language:** Slang, abbreviations (e.g., LOL, BRB), and non-standard grammar.

- **Unique Syntax:** The use of hashtags (#), mentions (@), and emojis adds semantic layers that traditional models may not understand.

- **Conciseness and Lack of Context:** Platforms like X (formerly Twitter) enforce character limits, leading to abbreviated messages where context must often be inferred.

- **Misspellings and Errors:** User-generated content is rife with typographical errors.

17.2.1 Advanced Preprocessing for Noisy Social Media Text

To overcome these challenges, a sophisticated preprocessing strategy is not merely a preliminary step but a core component of the analysis pipeline. The goal is to normalize the chaotic nature of user-generated content while preserving its rich semantic intent. This involves several specialized techniques (Sarker, 2017).

Emoji and Emoticon Handling: Simply removing emojis discards a vital layer of emotional and contextual information. A more effective strategy involves translating them into textual representations that a model can understand. For instance, a simple mapping can convert '♥' to the token 'love' and '😠' to 'angry.' More advanced approaches utilize specialized emoji embedding models, which learn vector representations for emojis based on the textual contexts in which they appear. This allows the model to understand that emojis like '😄' and '😂' are semantically closer to 'laughing' and 'joy' than they are to 'sadness.' This conversion of visual symbols into meaningful textual or vector-based features is critical for accurate sentiment and emotion analysis, as emojis often serve as the primary indicator of a user's tone (Eisner et al., 2016; Yuang et al., 2022).

Slang, Abbreviation, and Neologism Normalization: Social media is a crucible for new language. To handle the constant influx of slang (e.g., "spill the tea"), abbreviations ('iykyk' for "if you know, you know"), and neologisms, a static dictionary is insufficient. Effective SBI systems employ dynamic lexicons that are continuously updated. These can be built by scraping urban dictionaries or by using semi-supervised learning techniques to identify and define new terms from the data itself. Furthermore, the subword tokenization methods discussed in Chapter 4, such as byte-pair encoding (BPE) or WordPiece, are particularly powerful here. They can break down an unknown word like 'unbothered' into known subwords like 'un-,' 'bother,' and '-ed.' While 'unbothered' is a real, valid term found in standard dictionaries, it may not have been frequent enough to earn its own unique entry in the fixed vocabulary of a specific neural model, causing it to be treated as an out-of-vocabulary (OOV) word. This process, allowing the model to infer its meaning from its constituent parts, thus provides a robust defense against the out-of-vocabulary problem (Ahmed, 2015).

Strategic Handling of Mentions and Hashtags: Mentions (@username) and hashtags (#topic) are not just text; they are structured metadata that signifies connections and themes. A robust pipeline must treat them strategically. One approach is to replace all mentions of the specific user with a generic [USER_MENTION] token to reduce data sparsity when the user is not relevant to the analysis. Another is to strip the '@' symbol and treat the username as a proper noun for entity analysis. Hashtags can be segmented (e.g., #newlaptop -> new laptop) to be treated as regular text, or they can be preserved as unique tokens to be used as features for topic modeling, as they explicitly signal the primary subject of the content. The choice of strategy depends on the analytical goal: are we interested in the content of the message or the network of conversation? (Ariffin & Tiun, 2022).

17.3 Core NLP Applications in Social Media Analytics

SBI leverages a suite of advanced NLP tools to extract valuable insights from unstructured social media text. These tools enable a fine-grained analysis of public discourse, moving far beyond simple keyword tracking (Katya & Rahman, 2025).

17.3.1 Application: Monitoring Brand Sentiment and Emotion

Leveraging the sentiment analysis models discussed in Chapter 14, SBI's primary goal is to gauge public opinion about a brand, product, or event. This goes beyond a simple positive/negative binary to include more granular emotion detection, identifying feelings like joy, anger, surprise, or disgust. This real-time monitoring allows organizations to track brand health, measure the impact of marketing campaigns, and quickly respond to shifts in consumer attitude (Aruna et al., 2025).

17.3.2 Application: Granular Product Feedback with ABSA

While overall sentiment is useful, aspect-based sentiment analysis (ABSA) provides far more actionable intelligence. Instead of assigning a single sentiment score to a whole document, ABSA identifies specific *aspects* (features or topics) within the text and determines the sentiment for each one individually. For example, in the review, "The camera is incredible, but the battery life is a disappointment," ABSA would identify:

- camera -> positive

- battery life -> negative

This level of detail is invaluable for product development and targeted marketing (Hua et al., 2024).

Figure 17.1

Aspect-Based Sentiment Analysis (ABSA): Moving Beyond a Single Score to Granular Insights

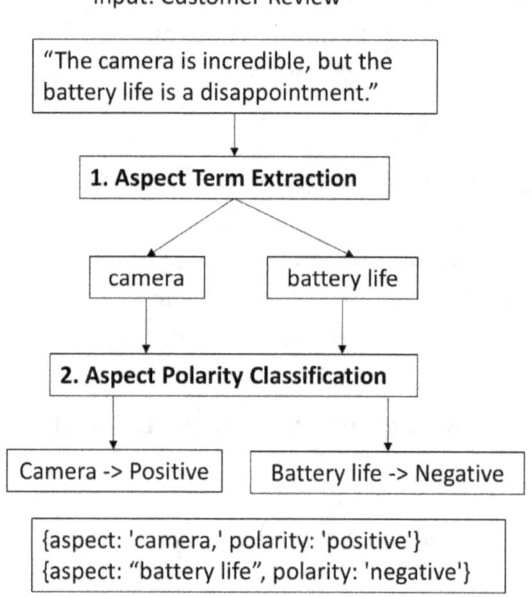

The Technical Mechanics of Aspect-Based Sentiment Analysis

Achieving this level of granularity requires breaking the problem down into a pipeline of two or three core sub-tasks, each of which is a challenging NLP problem in its own right (Shang et al., 2025).

1. **Aspect Term Extraction (ATE):** The first step is to identify the explicit features or aspects being discussed in the text. This is typically framed as a sequence labeling task, identical in form to the named entity recognition (NER) task detailed in Chapter 16. The model is trained to tag sequences of words that represent a product attribute. For example, in "The screen resolution is

amazing," the model would identify "screen resolution" as an aspect term. State-of-the-art approaches for ATE often use fine-tuned transformer-based models like BERT or specialized BiLSTM-CRF architectures, which are highly effective at learning the boundaries of multi-word aspect terms from annotated training data (Chauhan & Kumar, 2018).

2. **Aspect Polarity Classification (APC):** Once the aspect terms have been extracted, the next step is to determine the sentiment polarity (e.g., positive, negative, neutral) associated with each one. This is a classification task. The model is given the aspect term and its surrounding sentence as input and must predict the sentiment. For instance, given the aspect "battery life" and the sentence "The battery life is terrible," the model should classify the polarity as negative. This requires the model to understand the relationship between the aspect and the sentiment-bearing words (e.g., 'terrible') in the sentence (Schouten & Frasincar, 2016).

3. **Implicit Aspect Identification (Advanced):** The most challenging part of ABSA is handling implicit aspects, where the feature being discussed is not explicitly mentioned. In a sentence like, "This phone is so heavy," the implicit aspect is the phone's 'weight.' Or in "It costs a fortune," the implicit aspect is 'price.' Solving this requires a model with deeper world knowledge and the ability to perform commonsense reasoning. These systems often rely on large, pre-trained language models that have been fine-tuned on specialized datasets containing examples of implicit aspect-sentiment pairs. This remains an active and difficult area of research (Maitama et al., 2020; Azman et al., 2024).

17.3.3 Application: Discovering Conversation Themes with Topic Modeling

Sentiment analysis helps us understand how people feel, while topic modeling reveals what they are discussing. As mentioned in Chapter 14, topic modeling is an unsupervised learning technique that analyzes a large set of documents, such as thousands of tweets about a brand, and automatically groups them into clusters based on abstract 'topics.' By examining these prominent topics, organizations can identify emerging trends, detect product issues, or grasp key themes in discussions about competitors. Recent academic reviews highlight the crucial role of topic modeling in automatically detecting new subjects, tracking trends in public opinion, and generating actionable business intelligence from unstructured data (Albalawi et al., 2020).

17.3.4 Application: Identifying Key Entities in Conversations

Using the named entity recognition (NER) techniques detailed in Chapter 16, SBI platforms can automatically extract mentions of key entities from social media streams. This includes tracking mentions of company executives, competitor products, key industry influencers, and relevant locations. Given the noisy nature of social media, NER models for this domain must be robust to non-standard capitalization and spelling (Peres & Maheshwari, 2017).

17.4 Architecting an NLP-Powered SBI System

Building an industrial-strength SBI platform involves architecting a multi-stage data pipeline that is scalable, reliable, and efficient (Muntean, 2015; Talwandi et al., 2024).

1. **Data Ingestion:** Collecting data from social media platforms, news sites, and forums via APIs.

2. **Preprocessing:** Cleaning and normalizing the raw, noisy text.

3. **NLP Model Application:** Feeding the cleaned text into a suite of NLP models (e.g., for sentiment, topics, and named entities).

4. **Insight Aggregation:** Storing the structured outputs from the NLP models in a database.

5. **Visualization and Alerting:** Making the aggregated data accessible through customizable dashboards and real-time alerts.

17.4.1 Strategic and Technical Considerations in System Architecture

The high-level pipeline masks a series of critical engineering and strategic decisions that determine the system's cost, scalability, and real-time capabilities (Mositsa et al., 2023).

Batch Vs. Streaming Architectures: The choice of data processing architecture is fundamental. A **batch processing** system collects data over a period (e.g., 24 hours), processes it in a large batch, and updates the dashboards. This is computationally efficient and well-suited for generating daily or weekly reports on brand sentiment trends. However, it is inadequate for time-sensitive tasks like crisis management. For that, a **streaming architecture** is required. Using technologies like Apache Kafka for data ingestion and Apache Spark Streaming or Flink for processing, a streaming system analyzes social media posts in near real-time, enabling immediate alerts for sudden spikes in negative sentiment or the emergence of a viral issue. This real-time capability comes at a higher computational cost and architectural complexity (Marz & Warren, 2015).

Figure 17.2

Social Business Intelligence: Batch Vs. Streaming Architecture

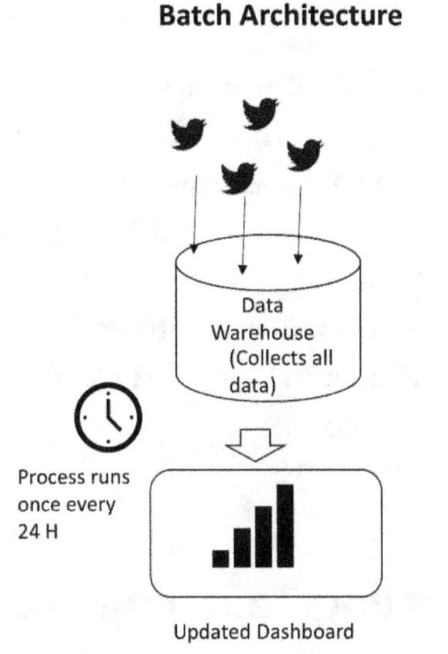

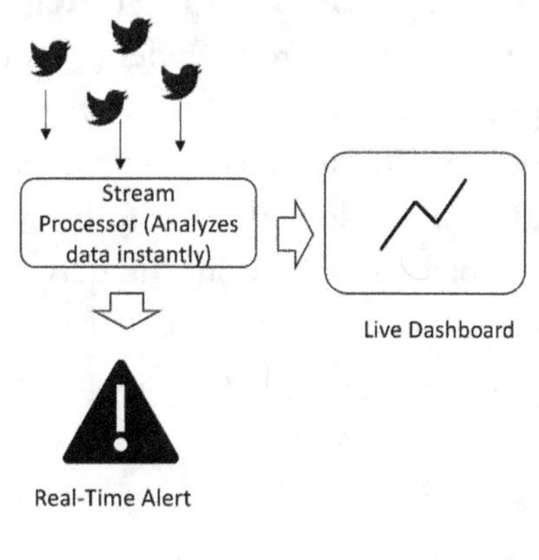

- **Pro**: Computationally efficient; good for historical trend reports
- **Con**: High latency; cannot react to real-time events
- **Use Case**: Weekly brand sentiment report

- **Pro**: Instant insights; enables real-time crisis response
- **Con**: More complex and computationally expensive
- **Use Case**: Detecting a viral negative tweet in real-time

The "Build vs. Buy" Decision: Organizations must decide whether to use third-party NLP APIs or build their own models.

- **Buy (API Services):** This approach involves utilizing pre-trained, managed natural language processing (NLP) services (such as those offered by major cloud and generative AI providers). These services offer powerful, out-of-the-box models for tasks like sentiment analysis and named entity recognition (NER). The primary advantages are fast implementation, minimal requirement for in-house machine learning expertise, and the transfer of model maintenance and scaling burden to the provider. The downsides are significant cost at scale, data privacy concerns (as data must be sent to a third party for processing), and a lack of customization for highly specialized, domain-specific language.

- **Build (Proprietary Models):** Building and fine-tuning models in-house using libraries like Hugging Face transformers provides maximum control and customization. This allows for the creation of highly accurate models tailored to a specific industry's jargon and context. It also ensures data remains within the organization's infrastructure. However, this path requires significant investment in ML engineering talent, computational resources (GPUs) for training, and ongoing effort for model maintenance and deployment (MLOps) (Alla & Adari, 2021).

Data Aggregation and Storage Strategy: The structured data extracted by the NLP models must be stored efficiently. A multi-database approach is often optimal. For example, sentiment scores over time are best stored in a **time-series database** like InfluxDB or Prometheus for fast querying and visualization of trends. Extracted entities and their relationships might be stored in a **graph database** like Neo4j, enabling complex queries about the network of influencers, brands, and topics. General metadata and aggregated statistics can be housed in a traditional **relational database** (like PostgreSQL) or a document store (like MongoDB) (Robinson et al., 2015).

17.5 The Multimodal Frontier: Integrating Text, Images, and Video

The analysis of social media is rapidly evolving beyond text alone. A significant portion of communication on platforms like Instagram, TikTok, and X (formerly Twitter) is visual. A post's meaning is often a synthesis of the text in the caption and the content of the accompanying image or video. A purely text-based analysis of a post that says "Loving my new laptop!" is incomplete if it misses the fact that the image shows a competitor's product. This has given rise to **multimodal social business intelligence**, which integrates computer vision techniques with NLP to create a more holistic understanding (Baltrušaitis et al., 2018).

This integrated analysis involves several layers. At a basic level, it can include object detection to identify products within an image, logo recognition to identify brands, and even visual sentiment analysis to infer emotion from facial expressions. The true power, however, comes from fusing these visual signals with the textual data. For example, a system could analyze a user's photo to identify that they are at a specific landmark (e.g., the Eiffel Tower) and combine that with the text "This vacation was a disaster" to provide highly specific negative feedback to a travel company (Soleymani et al., 2017).

This fusion of text and vision is powered by advanced multimodal AI models. Architectures based on **contrastive language-image pre-training (CLIP)** learn a shared embedding space where both images and text can be represented as vectors. This allows the model to understand the relationship between a photo of a dog and the text "a photo of a dog" by mapping them to nearby points in the vector space. This enables powerful zero-shot capabilities, such as searching a collection of images using natural-language queries without being explicitly trained on those specific image labels. For SBI, this means a system can analyze the sentiment of a post while simultaneously verifying that the image content aligns with the textual content, providing a much-needed layer of context and accuracy (Radford et al., 2021).

Figure 17.3

The CLIP Model: Creating a Joint Embedding Space for Text and Images

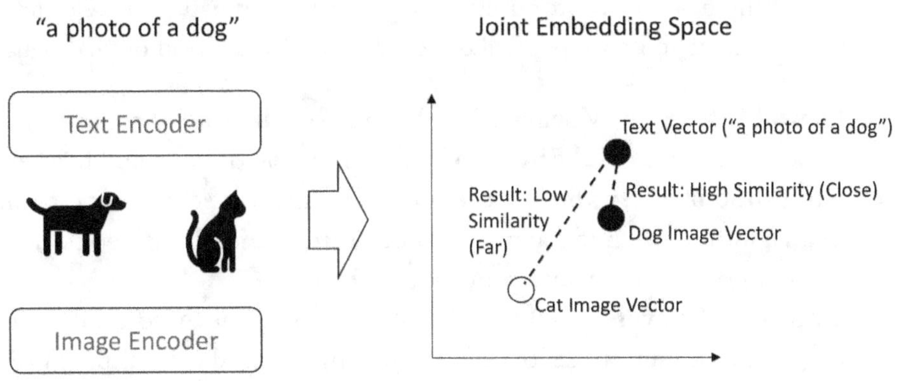

17.6 Ethical Challenges in Social NLP

The power to analyze population-level data at such a scale necessitates a rigorous ethical framework, as will be discussed in detail in Chapter 18. Key challenges specific to social NLP include:

- **Algorithmic Bias and Fairness:** Models trained on internet data can learn and perpetuate existing societal biases, potentially associating negative sentiment with specific demographic groups (Mehrabi et al., 2021).

- **Privacy:** While the data is public, users may not fully comprehend how their posts are being aggregated and analyzed for commercial purposes. The ethical implications of inferring sensitive attributes from public data are significant (Saxena et al., 2025; Hadi & Ibrahim, 2025).

- **Misinterpretation and Overreach:** Models can misinterpret sarcasm or cultural nuances, leading to flawed business decisions. The risk of using these insights for manipulative advertising or political targeting is a major concern (Tufekci, 2014; Sahakyan et al., 2025)

17.7 Conclusion and Future Trends

Social business intelligence (SBI) represents a powerful, real-world application of the entire suite of modern NLP techniques. By moving beyond simple keyword tracking to sophisticated semantic and multimodal analysis, these systems offer in-depth insights into customers' voices and the competitive landscape.

The future of the field holds promise for even greater sophistication. Contemporary research is advancing in areas that will define the next generation of SBI platforms:

- **Causal Inference:** Moving beyond mere **correlation** (what happened) to understand the **causal drivers** of shifts in public opinion and market trends.
- **Multimodal Integration:** Developing systems that seamlessly combine textual insights with visual information from images and video to derive more holistic and accurate contextual meaning.
- **Improved Conversational AI:** Developing more nuanced chatbots that can engage in meaningful dialogue with customers based on real-time insights from social listening.

As these technologies evolve, the demand for practitioners who understand both the underlying AI and the ethical imperatives of its application will only continue to grow.

17.8 References

Ahmed, B. (2015). Lexical normalisation of Twitter Data. *2015 Science and Information Conference (SAI)*, 326–328. https://doi.org/10.1109/SAI.2015.7237164

Albalawi, R., Yeap, T. H., & Benyoucef, M. (2020). Using topic modeling methods for short-text data: A comparative analysis. *Frontiers in Artificial Intelligence*, *3*, 42. https://doi.org/10.3389/frai.2020.00042

Alla, S., & Adari, S. K. (2021). *Beginning MLOps*. Apress. https://doi.org/10.1007/978-1-4842-6549-9

Ariffin, S. N. A. N., & Tiun, S. (2022). Improved POS tagging model for malay Twitter data based on machine learning algorithm. *International Journal of Advanced Computer Science & Applications*, *13*(7). https://doi.org/10.14569/IJACSA.2022.0130730

Aruna, T., Sumanth, A. V. S., Reddy, P. S., Mani, B. B., & Reddy, M. S. (2025). Real-time sentiment analysis of comments and reviews: A Kafka driven approach. *International Conference on Electronics, Materials Engineering and Nano-Technology (Online)*, 1–5. https://doi.org/10.1109/IEMENTech65115.2025.10959510

Azman Busst, M. M., Anbananthen, K. S. M., Kannan, S., & Kannan, R. (2024). Extracting explicit and implicit aspects using deep learning. *Emerging Science Journal*, *8*(1), 61–76. https://doi.org/10.28991/ESJ-2024-08-01-05

Baltrušaitis, T., Ahuja, C., & Morency, L. P. (2018). Multimodal machine learning: A survey and taxonomy. *IEEE Transactions on Pattern Analysis and Machine Intelligence*, *41*(2), 423–443. https://doi.org/10.1109/TPAMI.2018.2798607

Chauhan, G. S., & Kumar Meena, Y. (2018). Prominent aspect term extraction in aspect based sentiment analysis. *2018 3rd International Conference and Workshops on Recent Advances and Innovations in Engineering (ICRAIE)*, 1–6. https://doi.org/10.1109/ICRAIE.2018.8710408

Eisner, B., Rocktäschel, T., Augenstein, I., Bošnjak, M., & Riedel, S. (2016). *Emoji2vec: Learning emoji representations from their description*. https://arxiv.org/abs/1609.08359

Hadi, H., & Ibrahim, I. M. (2025). Advancements in NLP for social media analytics. International *Journal of Scientific World*, *11*(1), 167–177. https://doi.org/10.14419/y6d9jv30

Hua, Y. C., Denny, P., Wicker, J., & Taskova, K. (2024). A systematic review of aspect-based sentiment analysis: Domains, methods, and trends. *Artificial Intelligence Review*, *57*, 296. https://doi.org/10.1007/s10462-024-10906-z

Katya, E., & Rahman, S. (2025). Applications of natural language processing in social media sentiment analysis. *International Journal of Recent Advances in Engineering and Technology*, *13*(1), 11–15.

Li, C.-T., Ku, L.-W., Tsai, Y.-C., & Wang, W.-Y. (2022). SocialNLP'22: 10th International workshop on natural language processing for social media. In R. Troncy, L. Médini, I. Herman, & F. Laforest (Eds.), *Companion Proceedings of the Web Conference 2022* (pp. 849–851). ACM. https://doi.org/10.1145/3487553.3524876

Maitama, J. Z., Idris, N., Abdi, A., Shuib, L., & Fauzi, R. (2020). Systematic review on implicit and explicit aspect extraction in sentiment analysis. *IEEE Access*, *8*, 1–1. https://doi.org/10.1109/ACCESS.2020.3031217

Mehrabi, N., Morstatter, F., Saxena, N., Lerman, K., & Galstyan, A. (2021). A survey on bias and fairness in machine learning. *ACM Computing Surveys*, 54(6), 1–35. https://doi.org/10.1145/3457607

Marz, N., & Warren, J. (2015). *Big data: Principles and best practices of scalable real-time data systems*. Manning Publications Co

Mositsa, R. J., Van der Poll, J. A., & Dongmo, C. (2023). Towards a conceptual framework for data management in business intelligence. *Information (Basel)*, *14*(10), 547. https://doi.org/10.3390/info14100547

Muntean, M. (2015). Considerations regarding business intelligence in cloud context. *Informatica Economica*, *19*(4/2015), 55–67. https://doi.org/10.12948/issn14531305/19.4.2015.05

Peres, R., Esteves, D., & Maheshwari, G. (2017). Bidirectional LSTM with a context input window for named entity recognition in Tweets. *Proceedings of the Knowledge Capture Conference*, 1–4. https://doi.org/10.1145/3148011.3154478

Radford, A., Kim, J. W., Hallacy, C., Ramesh, A., Goh, G., Agarwal, S., Sastry, G., Askell, A., Mishkin, P., Clark, J., Krueger, G., & Sutskever, I. (2021). *Learning transferable visual models from natural language supervision*. arXiv. https://doi.org/10.48550/arXiv.2103.00020.

Robinson, I., Webber, J., & Eifrem, E. (2015). *Graph databases* (Second edition.). O'Reilly Media.

Sahakyan, H., Gevorgyan, A., & Malkjyan, A. (2025). From disciplinary societies to algorithmic control: Rethinking foucault's human subject in the digital age. *Philosophies (Basel)*, *10*(4), 73. https://doi.org/10.3390/philosophies10040073

Sarker, A. (2017). A customizable pipeline for social media text normalization. *Social Network Analysis and Mining*, *7*(1), 45. https://doi.org/10.1007/s13278-017-0464-z

Saxena, V., Tamò-Larrieux, A., Van Dijck, G., & Spanakis, G. (2025). Responsible guidelines for authorship attribution tasks in NLP: Responsible guidelines for authorship attribution tasks in NLP. *Ethics and Information Technology*, *27*(2), Article 16. https://doi.org/10.1007/s10676-025-09821-w

Schouten, K., & Frasincar, F. (2016). Survey on aspect-level sentiment analysis. *IEEE Transactions on Knowledge and Data Engineering*, *28*(3), 813–830. https://doi.org/10.1109/TKDE.2015.2485209

Shang, J., Zhang, Y., Zhong, L., & Li, R. (2025). Syntactic-enhanced multi-task learning model for aspect sentiment triplet extraction: Syntactic-enhanced multi-task learning model for aspect sentiment triplet extraction. *Data Science and Engineering*, *10*(3), 515–531. https://doi.org/10.1007/s41019-025-00289-8

Soleymani, M., Garcia, D., Jou, B., Schuller, B., Chang, S. F., & Pantic, M. (2017). A survey of multimodal sentiment analysis. *Image and Vision Computing*, *65*, 3–14.

Talwandi, N. S., Shah, A. K., Thakur, R., Khare, S., & Thakur, P. (2024). Cloud based data analytics for business intelligence. *2024 2nd International Conference on Advances in Computation, Communication and Information Technology (ICAICCIT)*, *1*, 1078–1081. https://doi.org/10.1109/ICAICCIT64383.2024.10912298

Tufekci, Z. (2014). Engineering the public: Big data, surveillance and computational politics. *First Monday*, *19*(7). https://doi.org/10.5210/fm.v19i7.4901

Yuan, X., Hu, J., Zhang, X., & Lv, H. (2022). Pay attention to emoji: Feature fusion network with emograph2vec model for sentiment analysis. *International Conference on Pattern Recognition*, 1529–1535. https://doi.org/10.1109/ICPR56361.2022.9956494

17.9 Glossary

Algorithmic Bias: A phenomenon where an algorithm produces systematically prejudiced results due to biased training data.

Apache Flink: A framework and distributed processing engine designed for low-latency, stateful stream processing. Unlike micro-batch systems, Flink handles data event-by-event, making it ideal for tasks in SBI that require immediate, millisecond-level reaction times, such as generating real-time alerts.

Apache Kafka: A highly scalable and durable distributed streaming platform used for building real-time data pipelines. In social business intelligence (SBI), it serves as the industry standard for high-throughput data ingestion, reliably storing and organizing the massive, incoming streams of social media data.

Apache Spark Streaming: An extension of the Apache Spark framework that enables scalable and fault-tolerant processing of data streams in near real-time. It achieves this by processing data in small, continuous batches (micro-batches).

Aspect-Based Sentiment Analysis (ABSA): A text analysis technique that identifies the sentiment associated with specific features or aspects of a product or service.

Batch Vs. Streaming Architectures: A fundamental choice in data processing. Batch processing analyzes large volumes of data at scheduled intervals, while streaming architecture analyzes data in near real-time as it arrives.

Causal Inference: A field of study focused on determining the cause-and-effect relationships between variables, moving beyond simple correlation.

CLIP (Contrastive Language-Image Pre-Training): An AI model that learns visual concepts from natural language supervision, enabling it to understand the relationship between images and text.

MLOps (Machine Learning Operations): A set of practices that combines machine learning (ML), DevOps, and data engineering to standardize and streamline the process of building, deploying, and maintaining ML models in production reliably and efficiently. It focuses on the automation of the entire machine learning lifecycle to ensure models remain accurate over time.

Multimodal Analysis: The analysis of data from multiple modalities, such as text, images, and video, to gain a more comprehensive understanding.

Social Business Intelligence (SBI): The practice of collecting and analyzing social media data to inform business strategy and decision-making.

Topic Modeling: An unsupervised machine learning technique used to discover abstract 'topics' that occur in a collection of documents.

17.10 Review Questions and Discussion Topics

1. **Preprocessing Challenges:** Propose a detailed preprocessing pipeline for tweets. Which steps would you take and in what order? How might you handle emojis and hashtags to retain their semantic value?

2. **Model Selection:** Compare and contrast the use of sentiment analysis versus aspect-based sentiment analysis (ABSA) for a company analyzing reviews of its new laptop. In what scenario would simple sentiment analysis be sufficient? When would the added complexity and cost of ABSA be justified?

3. **Ethical Dilemma - Bias:** An NLP model for your company's hiring platform analyzes the social media profiles of candidates to flag 'unprofessional' content. You discover the model is significantly more likely to flag posts written in African American Vernacular English (AAVE) as unprofessional. Based on the ethical principles discussed in the book, what are the potential harms and what steps could you take to mitigate this bias?

Chapter 18

Responsible AI: The Ethical and Societal Implications of NLP

18.1 Introduction: Beyond Accuracy—The New Imperative

Throughout this book, we have journeyed through the remarkable evolution of natural language processing, from the statistical foundations of N-grams to the architectural elegance of the transformer. We have learned how to build systems that can translate, summarize, classify, and generate language with unprecedented accuracy. Yet, as the capabilities of these models have grown, so too has their impact on society. The models we build are not deployed in a vacuum; they shape opinions, influence decisions in hiring and finance, and mediate our access to information. Therefore, the responsibility of the modern NLP practitioner extends far beyond optimizing for accuracy on a benchmark dataset. The new imperative is to build systems that are not only powerful but also fair, transparent, accountable, and safe. This final chapter confronts the complex ethical and societal dimensions of our work, providing a framework for responsible innovation and a call to action for the field's future.

18.2 A Framework for Responsible AI

To navigate the complex ethical landscape, the AI community has begun to converge on a set of core principles that should guide the development and deployment of NLP systems:

- **Fairness and Non-maleficence:** Systems should not perpetuate or amplify unjust societal biases, and they should be designed to avoid causing foreseeable harm to individuals or groups.

- **Transparency and Explainability:** The decision-making processes of AI systems should be transparent and understandable to both their users and developers. We should be able to ask why a model made a particular prediction.

- **Accountability and Governance:** There must be clear lines of responsibility for the outcomes of AI systems. This involves establishing robust governance structures, human oversight, and mechanisms for redress when things go wrong.

- **Privacy and Data Security:** Systems must respect user privacy, protect sensitive data, and be designed with data minimization principles in mind.

These principles are not merely abstract ideals; they represent concrete design goals that require dedicated technical solutions, rigorous testing, and a continuous commitment to ethical reflection throughout the entire AI lifecycle (Floridi et al., 2018).

Figure 18.1

The Four Pillars of Responsible AI

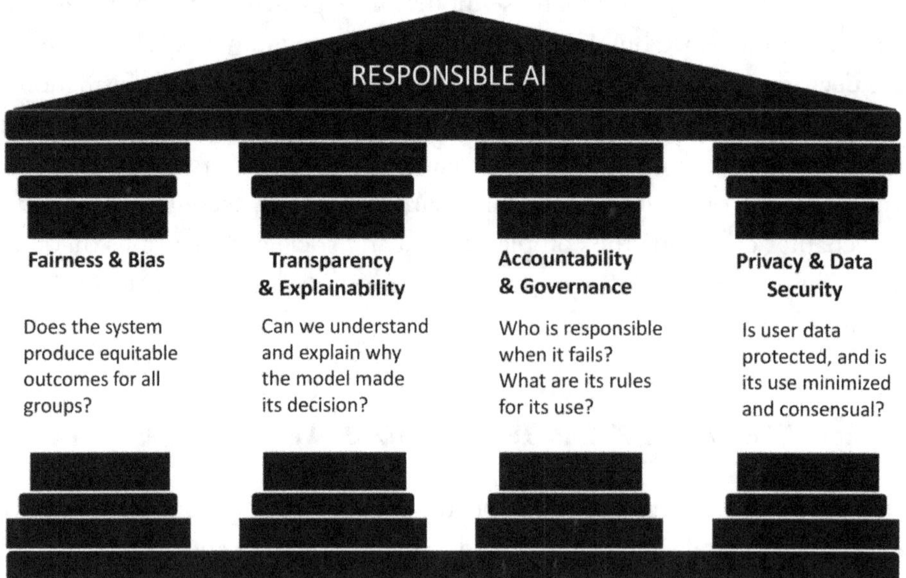

18.3 The Pervasive Challenge of Bias in NLP

Perhaps the most immediate and well-studied ethical challenge in NLP is bias. Models trained on vast corpora of human-generated text inevitably learn, and often amplify, the societal biases reflected in that data. This bias can manifest in several ways (Mehrabi et al., 2021):

- **Data Bias:** If historical data reflects that a certain demographic group has been underrepresented in a particular profession, a model trained on this data will learn that association. For example, word embeddings trained on general web text have been shown to produce analogies such as "man is to computer programmer as woman is to homemaker" (Bolukbasi et al., 2016; Zhang et al., 2020).

- **Algorithmic Bias:** The model's design can introduce bias. For example, an optimization function that aims to maximize overall accuracy might do so by sacrificing fairness for minority subgroups if they are underrepresented in the data. Consider a medical diagnostic model in which 95% of the training data come from demographic A and only 5% from demographic B. The algorithm can achieve a stellar 96% overall accuracy by optimizing exclusively for group A, even if its accuracy for the minority group B drops to a dangerous 50%. The math rewards the 'global' win while ignoring the 'local' failure.

- **Human Evaluation Bias:** Humans who label data or evaluate model outputs bring their own implicit biases to the task, which can be encoded into the system. For instance, in the task of detecting online toxicity or hate speech, human annotators who are unfamiliar with certain linguistic variations, such as African American English (AAE), may mistakenly flag non-offensive text containing AAE markers as toxic. The model, trained on these subjective labels, subsequently learns to associate the dialect itself—rather than the intent of the message—with a high probability of offensiveness, resulting in the unfair penalization of a specific demographic's speech (Sap et al., 2019; Okpala & Cheng, 2025).

18.3.1 Technical Approaches to Auditing and Mitigating Bias

Addressing bias is a critical and active area of research that has moved from identification to intervention. Key approaches can be categorized into auditing and mitigation (Bansal, 2022; Gallegos et al., 2024).

Auditing for Bias: Before bias can be fixed, it must be measured. A prominent technique for this is the **word embedding association test (WEAT)**, which quantifies associations between different word groups. Conceptually, it measures the similarity between two sets of target words (e.g., "male names" vs. "female names") and two sets of attribute words (e.g., "career words" vs. "family words"). By comparing cosine similarities, it can produce a score indicating, for instance, whether the embeddings show a stronger association between male names and career words than between female names and career words. A statistically significant score reveals a quantifiable gender bias encoded in the model's representations (Caliskan et al., 2017). While WEAT remains the standard for static models, modern research has adapted this framework into metrics like the **sentence encoder association test (SEAT)** and **CrowS-Pairs** to audit the more complex, contextual biases found in large language models (LLMs) (May et al., 2019; Nangia et al., 2020). Similar tests have been developed to audit models for racial, ethnic, and other forms of societal bias.

Figure 18.2

Three Points of Intervention for Mitigating Bias

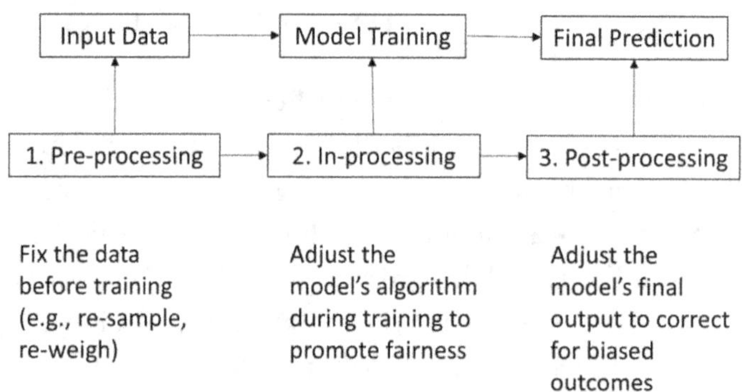

Mitigation Strategies: Once bias is identified, it can be addressed at different stages of the machine learning pipeline. **Pre-processing** techniques focus on the data itself, involving strategies like resampling to create a more balanced dataset or augmenting data for underrepresented groups. **In-processing** methods modify the model's training process directly. This can involve adding a "fairness penalty" to the model's loss function, which discourages the model from making predictions that disproportionately harm a specific demographic group. **Post-processing** techniques operate on the model's outputs. For example, if a loan application model is found to have a different approval threshold for different groups, its prediction scores can be adjusted after the fact to ensure equitable outcomes. Each approach has trade-offs between fairness, accuracy, and implementation complexity (Mehrabi et al., 2021).

18.4 Misinformation and the Dual-Use Nature of Generative Models

Generative models, particularly LLMs, are a quintessential dual-use technology. The same model that can be used to draft an eloquent email or assist a student with their homework can also be used to generate realistic-sounding but entirely false 'news' articles, propaganda, or personalized phishing emails at an unprecedented scale (Marijan & Pullen, 2023).

This challenge has two primary facets:

- **Disinformation at Scale:** The ability to automate the creation of plausible text poses a significant threat to our information ecosystem, making it increasingly difficult for individuals to distinguish fact from fiction (Zellers et al., 2019; Chizhov et al., 2025; Li et al., 2025).

- **The 'Hallucination' Problem:** Large language models have a tendency to 'hallucinate'—to generate confident, fluent, and grammatically perfect sentences that are factually incorrect or nonsensical. This is because they are optimized to predict the next plausible word rather than to access a knowledge base of verified facts. This makes them unreliable for high-stakes applications requiring factual accuracy without significant safeguards (Ji et al., 2022).

18.4.1 Grounding LLMs: Technical Solutions to Hallucination

The 'hallucination' problem is a direct consequence of a model's purely autoregressive nature. To combat this, a leading architectural pattern is **retrieval-augmented generation (RAG)**. A RAG system enhances a standard LLM by connecting it to an external, trusted knowledge source. The process works in two stages: First, when a user provides a prompt, a 'retriever' component (often a specialized search model) queries a knowledge base—such as a company's internal wiki, a set of legal documents, or a curated database of facts—and pulls the most relevant snippets of information. Second, these retrieved snippets are combined with the original prompt and fed into the LLM 'generator' with an explicit instruction to synthesize its answer solely from the provided context. This grounds the model's response in verifiable external facts, dramatically reducing the likelihood of hallucinations and allowing the system to provide up-to-date information without requiring complete retraining (Lewis et al., 2020; Su et al., 2025; Tang et al., 2025). Other techniques include training secondary models to fact-check an LLM's output or developing methods for models to express uncertainty in their responses.

18.5 Privacy, Data Governance, and Inference

Modern NLP models are trained on massive datasets, much of which is scraped from the public web and may contain personally identifiable information (PII). This raises significant privacy risks (Bender et al., 2021). Even if explicit PII, such as names and addresses, is removed, models can inadvertently memorize and later regurgitate sensitive information present in their training data. Furthermore, the unique 'fingerprint' or 'stylome' of an individual's writing can sometimes be used to re-identify them across different platforms.

18.5.1 Technical Approaches to Privacy Preservation

Addressing these issues requires a "privacy-by-design" approach that incorporates techniques from privacy-preserving machine learning (PPML). One of the most important concepts is **differential privacy**, which offers a formal, mathematical guarantee of privacy. The core intuition is to inject a carefully calibrated amount of statistical noise into the training data, the model's parameters, or the query results. This noise is just large enough to make it mathematically impossible to determine whether any single individual's data was included in the training set, thus protecting their privacy, while being small enough to not significantly degrade the overall accuracy of the model (Dwork, 2008; Li et al., 2022; Tianqing et al., 2017). A complementary approach is **federated learning**, an architectural pattern for training models on decentralized data. Instead of collecting all data on a central server, the model is sent to the user's local device (e.g., a mobile phone) to be trained on their data. Only the resulting model updates, not the raw data, are sent back to be aggregated into a global model. This ensures that sensitive, raw data never leaves the user's control (McMahan et al., 2017; Wu et al., 2022).

18.6 Transparency and Explainable AI (XAI)

Many of the most powerful deep learning models function as "black boxes." We know they work, but we often don't know why they make the decisions they do. This lack of transparency is a major barrier to trust and accountability, especially in critical domains. Explainable AI (XAI) is an emerging field dedicated to developing methods to peer inside these models (Gunning et al., 2019).

Techniques in NLP include:

- **Attention Visualization:** For attention-based models, visualizing the attention weights can show which parts of the input text the model "paid attention to" when making a prediction. These visualizations typically take the form of heatmaps, using varying color intensity to indicate the strength of the attention weight assigned to each word. This technique is based on the intuitive idea that higher focus should correlate with higher contribution to the output.

- **LIME (Local Interpretable Model-Agnostic Explanations):** A model-agnostic technique designed to explain the prediction of *any* black-box classifier, regardless of its underlying architecture (e.g., neural network, random forest)

18.6.1 Operationalizing Explainability

To move from theory to practice, practitioners can use several established XAI techniques. A popular method is **LIME (Local Interpretable Model-agnostic Explanations)**. To explain why a complex model made a specific prediction for a single instance (e.g., classifying an email as spam), LIME constructs a temporary, simpler model that is designed to explain only that one decision. It generates thousands of slight variations of the original input (e.g., by removing some words from the email) and gets predictions for each from the complex model. It then trains a simple, interpretable model, like a linear regression, on this new dataset of variations and their predictions. The weights of this simple model then provide a local explanation, highlighting which words were most important for the original 'spam' classification (Ribeiro et al., 2016; Parisineni & Pal, 2024).

Another powerful technique is **SHAP (SHapley Additive exPlanations)**. Grounded in cooperative game theory, SHAP calculates the **marginal contribution**—the **incremental impact**—of adding each feature to the model's decision. Conceptually, it asks: "How much does the prediction change *specifically because* this feature was included?" It does this by testing the feature in combination with every other possible subset of features to ensure a fair distribution of credit (Lundberg & Lee, 2017; Wu, 2025). This provides a more mathematically rigorous way to assign importance scores, ensuring that the explanations are consistent and locally accurate. While computationally more intensive than LIME, SHAP values have become a standard for robust feature attribution in high-stakes applications. It is also crucial to note the limitations of some methods. For instance, while attention visualizations are intuitive, a significant body of research argues that attention weights do not always provide a faithful explanation of a model's reasoning and can sometimes be misleading (Jain & Wallace, 2019).

Figure 18.3

Explainable AI (XAI): Opening the Black Box

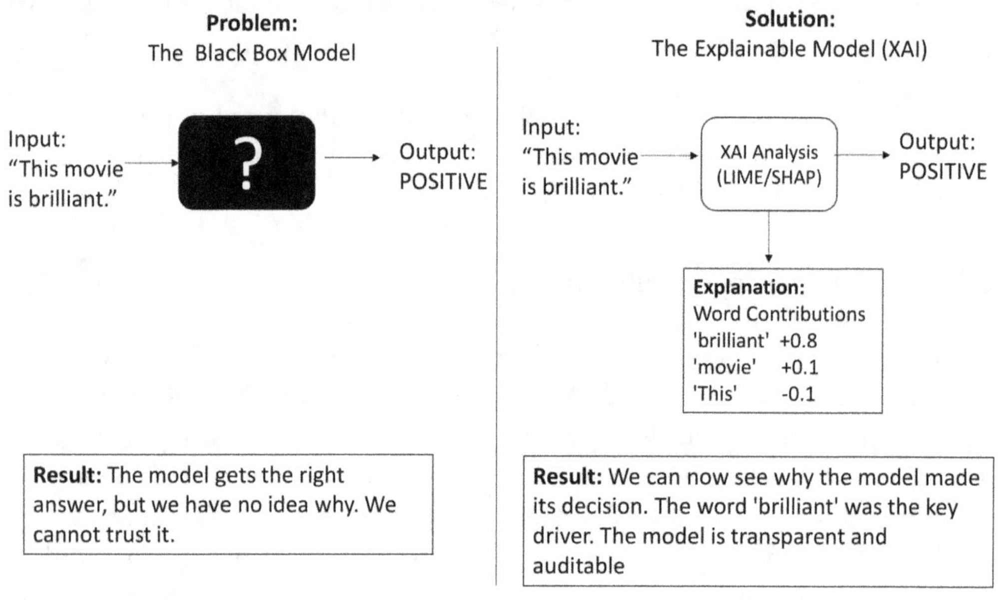

18.7 The Environmental Cost of Scale

A final, often overlooked, ethical consideration is the environmental impact of training large-scale NLP models. The computational resources required to train a state-of-the-art model can be immense, consuming vast amounts of electricity and contributing to a significant carbon footprint. The trend identified by early work in this area (Strubell et al., 2019; Różycki et al., 2025) has only accelerated, with subsequent analyses confirming that the carbon cost of training a single large model can be many times that of a round-trip trans-American flight (Patterson et al., 2021; Mitu & Mitu, 2024).

This raises important questions about sustainability and equity, as only a handful of large corporations and well-funded research labs have the resources to train these models from scratch. This has led to a call within the research community for a shift towards "Green AI"—a focus on developing models and methods that are computationally efficient, not just accurate. This includes prioritizing research on Green AI strategies: designing smaller, more efficient architectures that achieve high performance with fewer parameters; employing model distillation, where a compact 'student' model learns to mimic the behavior of a massive 'teacher' model at a fraction of the computational cost; and leveraging hardware acceleration through specialized chips optimized to execute AI operations with maximal energy efficiency (Schwartz et al., 2020; Barbierato & Gatti, 2024).

18.8 Conclusion: A Call for Responsible Innovation

Natural language processing has endowed us with capabilities that were the stuff of science fiction only a decade ago. However, the deployment of such transformative technology imposes a commensurate ethical obligation. The challenges of bias, misinformation, privacy, and transparency are not peripheral issues to be solved by others; they are central engineering and research problems that must be addressed with the same rigor and creativity we apply to model building. The future of NLP will be defined not just by the accuracy of our models, but by their impact on people and society. The goal is not simply to build machines that can understand language, but to build them in a way that is worthy of our trust.

18.9 References

Bansal, R. (2022). *A survey on bias and fairness in natural language processing.* https://doi.org/10.48550/arxiv.2204.09591

Barbierato, E., & Gatti, A. (2024). Toward green AI: A methodological survey of the scientific literature. *IEEE Access, 12,* 23989–24013. https://doi.org/10.1109/ACCESS.2024.3360705

Bender, E. M., Gebru, T., McMillan-Major, A., & Shmitchell, S. (2021). On the dangers of stochastic parrots: Can language models be too big? *Proceedings of the 2021 ACM Conference on Fairness, Accountability, and Transparency,* 610–623. https://doi.org/10.1145/3442188.3445922

Bolukbasi, T., Chang, K. W., Zou, J. Y., Saligrama, V., & Kalai, A. T. (2016). Man is to computer programmer as woman is to homemaker? Debiasing word embeddings. *Advances in Neural Information Processing Systems, 29,* 4349–4357.

Caliskan, A., Bryson, J. J., & Narayanan, A. (2017). Semantics derived automatically from language corpora contain human-like biases. *Science, 356*(6334), 183–186. https://doi.org/10.1126/science.aal4230

Chizhov, P., Nee, M., Langlais, P.-C., & Yamshchikov, I. P. (2025). What the HellaSwag? On the validity of common-sense reasoning benchmarks. https://doi.org/10.48550/arxiv.2504.07825

Dwork, C. (2008). Differential privacy: A survey of results. In M. Agrawal, D. Du, Z. Duan, & A. Li (Eds.), Theory and applications of models of computation (pp. 1–19). Springer. https://doi.org/10.1007/978-3-540-79228-4_1

Floridi, L., Cowls, J., Beltrametti, M., Chatila, R., Chazerand, P., Dignum, V., ... & Vayena, E. (2018). AI4People–An ethical framework for a good AI society: Opportunities, risks, principles, and recommendations. *Minds and Machines, 28*(4), 689–707. https://doi.org/10.1007/s11023-018-9482-5

Gallegos, I. O., Rossi, R. A., Barrow, J., Tanjim, M. M., Kim, S., Dernoncourt, F., Yu, T., Zhang, R., & Ahmed, N. K. (2024). Bias and fairness in large language models: A survey. *Computational Linguistics - Association for Computational Linguistics, 50*(3), 1097–1179. https://doi.org/10.1162/coli_a_00524

Gunning, D., Stefik, M., Choi, J., Miller, T., Stumpf, S., & Yang, G. Z. (2019). XAI–Explainable artificial intelligence. *Science Robotics, 4*(37), eaay7120. https://doi.org/10.1126/scirobotics.aay7120

Jain, S., & Wallace, B. C. (2019). Attention is not explanation. *Proceedings of the 2019 Conference of the North American Chapter of the Association for Computational Linguistics: Human Language Technologies,* Volume 1 (Long and Short Papers), 3543–3556. https://doi.org/10.18653/v1/N19-1357

Ji, S., Pan, S., Cambria, E., Marttinen, P., & Yu, P. S. (2022). A survey on knowledge graphs: Representation, acquisition, and applications. *IEEE Transactions on Neural Networks and Learning Systems, 33*(2), 494–514. https://doi.org/10.1109/TNNLS.2021.3070843

Lewis, P., Perez, E., Piktus, A., Petroni, F., Karpukhin, V., Goyal, N., Küttler, H., Lewis, M., Yih, W.-T., Rocktäschel, T., Riedel, S., & Kiela, D. (2020). *Retrieval-augmented generation for knowledge-intensive NLP tasks.* arXiv. https://doi.org/10.48550/arXiv.2005.11401.

Li, X., Li, M., Men, R., Zhang, Y., Bao, K., Wang, W., Feng, F., Liu, D., & Lin, J. (2025). *HellaSwag-Pro: A large-scale bilingual benchmark for evaluating the robustness of llms in commonsense reasoning.* https://doi.org/10.48550/arxiv.2502.11393

Li, N., Lyu, M., Su, D., & Yang, W. (2022). *Differential privacy: From theory to practice* (M. Lyu, W. Yang, & D. Su, Eds.; 1st ed.). Springer Nature. https://doi.org/10.1007/978-3-031-02350-7

Lundberg, S. M., & Lee, S.-I. (2017). *A unified approach to interpreting model predictions.* arXiv. https://doi.org/10.48550/arXiv.1705.07874.

Marijan, B., & Pullen, R. (2023). The dilemma of dual-use AI. *Ploughshares Monitor, 44*(3), 14.

May, C., Wang, A., Bordia, S., Bowman, S. R., & Rudinger, R. (2019). On measuring social biases in sentence encoders. *Proceedings of the 2019 Conference of the North American Chapter of the Association for Computational Linguistics: Human Language Technologies,* Volume 1 (Long and Short Papers), 622–628. https://doi.org/10.18653/v1/N19-1063

McMahan, B., Moore, E., Ramage, D., Hampson, S., & y Arcas, B. A. (2017). Communication-efficient learning of deep networks from decentralized data. *Proceedings of the 20th International Conference on Artificial Intelligence and Statistics, 54,* 1273–1282.

Mehrabi, N., Morstatter, F., Saxena, N., Lerman, K., & Galstyan, A. (2021). A survey on bias and fairness in machine learning. *ACM Computing Surveys, 54*(6), 1–35. https://doi.org/10.1145/3457607

Mitu, N. E., & Mitu, G. T. (2024). The hidden cost of AI: Carbon footprint and mitigation strategies. *Revista de Stiinte Politice, 84*, 9–16.

Nangia, N., Vania, C., Bhalerao, R., & Bowman, S. R. (2020). CrowS-Pairs: A challenge dataset for measuring social biases in masked language models. *Proceedings of the 2020 Conference on Empirical Methods in Natural Language Processing (EMNLP)*, 1953–1967. https://doi.org/10.18653/v1/2020.emnlp-main.154

Okpala, E., & Cheng, L. (2025). Large language model annotation bias in hate speech detection. *Proceedings of the International AAAI Conference on Web and Social Media, 19*, 1389–1418. https://doi.org/10.1609/icwsm.v19i1.35879

Parisineni, S. R. A., & Pal, M. (2024). Enhancing trust and interpretability of complex machine learning models using local interpretable model agnostic shap explanations. *International Journal of Data Science and Analytics, 18*(4), 457–466. https://doi.org/10.1007/s41060-023-00458-w

Patterson, D., Gonzalez, J., Le, Q., Liang, C., Munguia, L. M., Rothchild, D., So, D., Texier, M., & Dean, J. (2021). *Carbon emissions and large neural network training.* arXiv. https://arxiv.org/abs/2104.10350

Ribeiro, M. T., Singh, S., & Guestrin, C. (2016). "Why should I trust you?": Explaining the predictions of any classifier. *Proceedings of the 22nd ACM SIGKDD International Conference on Knowledge Discovery and Data Mining*, 1135–1144. https://doi.org/10.1145/2939672.2939778

Różycki, R., Solarska, D. A., & Waligóra, G. (2025). Energy-aware machine learning models—a review of recent techniques and perspectives. *Energies (Basel), 18*(11), 2810. https://doi.org/10.3390/en18112810

Sap, M., Card, D., Gabriel, S., Choi, Y., & Smith, N. A. (2019). The risk of racial bias in hate speech detection. *Proceedings of the 57th Annual Meeting of the Association for Computational Linguistics*, 1668–1678. https://doi.org/10.18653/v1/P19-1163

Schwartz, R., Dodge, J., Smith, N. A., & Etzioni, O. (2020). Green AI. *Communications of the ACM, 63*(12), 54–63. https://doi.org/10.1145/3381831

Strubell, E., Ganesh, A., & McCallum, A. (2019). Energy and policy considerations for deep learning in NLP. *Proceedings of the 57th Annual Meeting of the Association for Computational Linguistics*, 3645–3650. https://doi.org/10.18653/v1/P19-1355

Su, W., Ai, Q., Zhan, J., Dong, Q., & Liu, Y. (2025). Dynamic and parametric retrieval-augmented generation. *Proceedings of the 48th International ACM SIGIR Conference on Research and Development in Information Retrieval*, 4118–4121. https://doi.org/10.1145/3726302.3731692

Tang, Y., Zhang, R., Guo, J., de Rijke, M., Fan, Y., & Cheng, X. (2025). Boosting retrieval-augmented generation with generation-augmented retrieval: A co-training approach. *Proceedings of the 48th International ACM SIGIR Conference on Research and Development in Information Retrieval*, 2441–2451. https://doi.org/10.1145/3726302.3729907

Tianqing Zhu, Gang Li, Wanlei Zhou, & Yu, P. S. (2017). Differentially private data publishing and analysis: A survey. *IEEE Transactions on Knowledge and Data Engineering, 29*(8), 1619–1638. https://doi.org/10.1109/TKDE.2017.2697856

Wu, L. (2025). A review of the transition from Shapley values and SHAP values to RGE. *Statistics (Berlin, DDR), 59*(5), 1161–1183. https://doi.org/10.1080/02331888.2025.2487853

Wu, C., Wu, F., Lyu, L., Huang, Y., & Xie, X. (2022). Communication-efficient federated learning via knowledge distillation. *Nature Communications, 13*(1), Article 2032. https://doi.org/10.1038/s41467-022-29763-x

Zhang, H., Sneyd, A., & Stevenson, M. (2020). Robustness and reliability of gender bias assessment in word embeddings: The role of base pairs. *In Proceedings of the 1st Conference of the Asia-Pacific Chapter of the Association for Computational Linguistics and the 10th International Joint Conference on Natural Language Processing* (pp. 759–769). Association for Computational Linguistics

Zellers, R., Holtzman, A., Bisk, Y., Farhadi, A., & Choi, Y. (2019). HellaSwag: Can a machine really finish your sentence? *Proceedings of the 57th Annual Meeting of the Association for Computational Linguistics*, 4791–4800. https://doi.org/10.18653/v1/P19-1472

18.10 Glossary

Algorithmic Bias: Systematic and repeatable errors in a computer system that create unfair outcomes, such as privileging one arbitrary group of users over others.

CrowS-Pairs (Crowdsourced Stereotype Pairs): A benchmark dataset and metric used to measure social bias in Masked Language Models. It uses pairs of sentences that are identical except for the demographic group mentioned (e.g., one stereotyping, one anti-stereotyping) to test which sentence the model assigns a higher probability to.

Differential Privacy: A system for publicly sharing information about a dataset by describing the patterns of groups within the dataset while withholding information about individuals.

Disinformation: False information that is deliberately spread to deceive people.

Dual-Use Technology: A technology that can be used for both peaceful and malicious purposes.

Explainable AI (XAI): A field of AI research that aims to produce models whose decisions and predictions can be understood by humans.

Fairness: In AI, the principle that a model's predictions should not be biased based on sensitive demographic attributes.

Federated Learning: A decentralized machine learning approach where a model is trained on user devices (e.g., mobile phones) without the raw data leaving the device, enhancing privacy.

Hallucination: A phenomenon where a large language model generates text that is plausible and grammatically correct but is factually incorrect or nonsensical.

LIME (Local Interpretable Model-agnostic Explanations): A technique that explains the predictions of any classifier in an interpretable and faithful manner by learning an interpretable model locally around the prediction.

Personally Identifiable Information (PII): Any data that could potentially be used to identify a specific individual.

Retrieval-Augmented Generation (RAG): An architecture that combines a generative LLM with an external information retrieval system to ground responses in specific, retrieved data, reducing hallucinations.

SEAT (Sentence Encoder Association Test): An adaptation of the WEAT method designed for contextual word embeddings (like BERT) and large language models. Instead of isolated words, it uses sentence templates (e.g., "This is a [career/family] word") to measure bias in full-sentence contexts.

SHAP (SHapley Additive exPlanations): A game-theoretic approach to explain the output of any machine learning model by assigning each feature an importance value for a particular prediction.

WEAT (Word Embedding Association Test): A statistical test used to measure bias in static word embeddings (like Word2Vec) by comparing the mathematical association between two sets of target words (e.g., male/female names) and two sets of attribute words (e.g., career/family terms).

18.11 Review Questions and Discussion Topics

1. **Bias in Word Embeddings:** The classic example vector('Man') - vector('Woman') is often shown to be analogous to vector('King') - vector('Queen') but also vector("Computer Programmer") - vector('Homemaker'). Discuss the societal implications of using these biased embeddings in a resume screening tool.

2. **The "Black Box" Problem:** You have deployed a sentiment analysis model to automatically flag toxic comments on a social media platform. The model works well, but occasionally flags innocuous comments as toxic. Why is it critical to have an XAI method in this scenario? How would being able to explain the model's error help you improve the system and build user trust?

3. **Generative AI and Academia:** Discuss the ethical implications of students using large language models to write essays. Is this plagiarism? How does it differ from using a calculator for math or a spell checker for writing? Propose a policy that a university could adopt to govern the use of these tools.

4. **Privacy Vs. Utility:** A healthcare company wants to use NLP to analyze patient therapy notes to identify individuals at high risk of self-harm. This could save lives, but it involves processing extremely sensitive data. What ethical principles from this chapter are in tension here? What safeguards (technical and procedural) would you demand be in place before even considering such a project?

5. **Your Role as an Ethical Practitioner:** Imagine you are a junior NLP engineer at a startup. You notice that the dataset being used to train a new product feature

is heavily skewed and underrepresents minority populations. Your manager is on a tight deadline and tells you to proceed anyway, saying, "We can fix it later." What are your ethical obligations in this situation? What steps could you take?

6. **The Environmental Cost of Scale:** Section 18.7 discusses the significant environmental impact of training large-scale NLP models. This creates a potential tension between the push for ever-more-powerful models and the principle of sustainability. Discuss the ethical implications of this "AI carbon footprint." Who bears the responsibility for this cost—researchers, corporations, or consumers of AI services? What potential solutions or changes in research focus could help mitigate this problem?

Appendices

Appendix A

A Technical Deep Dive into Preprocessing

Abstract: A Synthesis of Evolution and Impact

This appendix presents a thorough and comprehensive analysis of the historical evolution and ongoing computational significance of core natural language processing (NLP) pre-processing techniques. It delves into the development of tokenization, stemming, lemmatization, stop word removal, and the bag-of-words (BoW) model, tracing their origins and transformations over time. The analysis focuses on examining the profound impact these techniques have on computational processing, clarifying their modern applications, and providing clear insights into the rationale behind their use in contemporary NLP pipelines. A central theme of the appendix is the relentless pursuit of greater linguistic accuracy balanced against the demands of computational efficiency. It meticulously highlights the inherent trade-offs in various pre-processing approaches and explores the technological advancements that have shaped them, from simple heuristic-based methods to sophisticated, data-driven algorithms.

A.1. Introduction to Natural Language Processing Pre-processing

A.1.1. The Foundational Role of Pre-processing in NLP

Natural language processing (NLP) is a pivotal subfield of artificial intelligence dedicated to equipping machines with the ability to understand, interpret, and even generate human language in a way that is both meaningful and practically applicable (Jurafsky & Martin, 2023). This appendix provides a detailed computational analysis of the core pre-processing techniques introduced conceptually in Chapter 4, offering a deeper look at the algorithms and trade-offs that underpin this foundational stage of NLP. Within the intricate architecture of any NLP pipeline, pre-processing constitutes the indispensable first stage. This initial phase is dedicated to the critical tasks of cleaning, structuring, and preparing raw, unstructured text data for the subsequent, more complex stages of analysis (Manning et al., 2008).

The importance of this transformative step cannot be overstated. Pre-processing is the fundamental mechanism that converts human-readable language—with all its nuances, ambiguities, and inconsistencies—into a structured, machine-readable format. This structured representation is the bedrock upon which all further computational operations are built (Vijayarani et al., 2015). The overall robustness and reliability of an entire NLP system are therefore critically dependent on the quality and thoroughness of this foundational pre-processing phase.

A.1.2. Historical Context of NLP's Evolution

The historical trajectory of NLP is marked by a continuous quest for more effective and sophisticated methods to process and understand human language (Jurafsky & Martin, 2023). The initial phase of NLP was characterized by symbolic, rule-based systems that relied on manually crafted linguistic knowledge. A pivotal paradigm shift occurred in the 1980s with the advent of statistical NLP. This transformation was fueled by two key factors: increasing computational power and the growing availability of large-scale text corpora. These advancements enabled systems to learn language patterns directly from data, leading to the development of more scalable, adaptive, and robust models (Manning & Schütze, 1999).

The 1990s witnessed accelerated growth, with concepts like word sense disambiguation and information extraction gaining prominence. The 2000s and 2010s marked another revolutionary phase, driven by the integration of machine learning, particularly deep learning and neural networks. Post-2010, technologies like recurrent neural networks (RNNs), long short-term memory (LSTM) networks, and especially transformer-based architectures (e.g., BERT, GPT) propelled NLP into an era of unprecedented innovation, enabling machines to achieve near-human performance on a wide range of language tasks (Devlin et al., 2019).

A.2. Tokenization: From Simple Splits to Subword Units

A.2.1. Evolution of Tokenization Techniques

Tokenization is the quintessential first step in any NLP pre-processing pipeline. It is the process of breaking down a raw stream of text into smaller, discrete, and manageable units known as 'tokens' (Jurafsky & Martin, 2023). These tokens serve as the fundamental building blocks for all subsequent analysis. The granularity of these tokens can vary, ranging from entire words and subwords to individual characters. The ultimate goal of tokenization is to convert unstructured text into a structured sequence of tokens that can be readily processed by a machine (Manning et al., 2008).

Early techniques were characteristically simple, often relying on basic rules. The most common approach, **word tokenization**, involves segmenting text based on whitespace and punctuation. While computationally efficient, this method proved inadequate for handling complex linguistic phenomena such as contractions (e.g., 'don't'), hyphenated words (e.g., 'state-of-the-art'), or languages without explicit word delimiters (Gimpel et al., 2011).

The limitations of word-level tokenization, particularly the out-of-vocabulary (OOV) problem, spurred the development of **subword tokenization**. This approach breaks words into smaller, meaningful parts, such as roots, prefixes, and suffixes. Several key subword algorithms have become standard in modern NLP:

- **Byte-Pair Encoding (BPE):** Originally a data compression algorithm, BPE was adapted for NLP to address the rare word problem. The training process is iterative:
    1. Initialize the vocabulary with all individual characters present in the training corpus.
    2. Count the frequency of all adjacent token pairs.
    3. Identify the most frequent pair (e.g., 'e' and 's').
    4. Merge this pair into a new, single token (e.g., 'es') and add it to the vocabulary.
    5. Repeat steps 2-4 until the vocabulary reaches a predefined size (e.g., 30,000 merges). This process allows the model to represent any word, even OOV words, by composing them from known subword units (Sennrich et al., 2016).
- **WordPiece:** Employed by influential models like BERT, WordPiece is similar

to BPE but uses a different merge criterion. Instead of merging the most frequent pair, it merges the pair that maximizes the likelihood of the training data.

This criterion is calculated as the ratio of the merged token's probability to the probabilities of its two constituent parts:

$$P(xy) / (P(x) \times P(y))$$

A high score means the pair x followed by y occurs much more frequently than one would expect if x and y appeared independently. This method tends to favor merges that are 'surprisingly' common, rather than just frequent, often yielding a more efficient vocabulary (Schuster & Nakajima, 2012).

- **SentencePiece:** This is a language-agnostic tokenization toolkit that implements both BPE and another algorithm called **unigram language model** tokenization. Its key innovation is treating the input text as a raw stream of Unicode characters, directly encoding whitespace as a special token. This makes it highly effective for multilingual models and removes the need for language-specific pre-tokenization rules (Kudo & Richardson, 2018).

A.2.2. Impact on Computational Processing

The evolution of tokenization has been heavily influenced by the persistent challenge of handling OOV words. Simple word-level tokenization, while fast, resulted in very large vocabularies and was unable to represent words not seen in the training corpus (Sennrich et al., 2016). This directly spurred the development of subword tokenization, which was explicitly designed to decompose rare and unknown words into known sub-units. However, this advancement introduced a computational trade-off. Subword tokenization algorithms are more complex and computationally expensive than simple splitting rules. They require an initial training phase to build the vocabulary and careful tuning of hyperparameters like vocabulary size (Kudo & Richardson, 2018). This increased computational demand is the price paid for solving the OOV problem and enabling models to handle morphologically rich languages more effectively.

A.2.3. Current Applications and Usage in Modern NLP

Tokenization is a universal prerequisite for virtually all NLP tasks. Its method is chosen based on the model architecture and task requirements:

- **Machine Translation:** Tokenization defines the linguistic units that are mapped from a source language to a target language. Subword tokenization is particularly effective here for handling rare words and aligning phrases across languages (Sennrich et al., 2016).
- **Information Retrieval and Search Engines:** Search engines tokenize user queries and web documents to create indexes, enabling accurate matching and retrieval (Manning et al., 2008).
- **Large Language Models (LLMs):** For LLMs, subword tokenization is the fundamental first step. Input text is broken into subword units, which are then converted to numerical IDs and fed into the model's embedding and transformer layers (Devlin et al., 2019).

Despite its advancements, tokenization still faces challenges, including handling linguistic ambiguity, processing languages without clear word boundaries, and dealing with noisy, unstructured text (Jurafsky & Martin, 2023).

A.3. Stemming: Reducing Words to Their Roots

A.3.1. Historical Development of Stemming Algorithms

Stemming is a text normalization technique that reduces words to their root or base form—the 'stem'—by heuristically removing prefixes and suffixes. One of the earliest rule-based stemmers was developed by Julie Beth Lovins in 1968 (Lovins, 1968).

The most widely recognized and influential stemmer is the **Porter stemmer**, created by Martin Porter in 1980. It became a de facto standard due to its simplicity and effectiveness for English. The Porter stemmer uses a five-step, rule-based algorithm that iteratively applies suffix-stripping rules based on the measure of the word stem (the number of vowel-consonant sequences) (Porter, 1980). Porter later developed the **Snowball** framework, which includes an improved English stemmer (Porter2) and stemmers for numerous other languages.

A.3.2. Computational Implications and Limitations

Stemming offers significant computational advantages. Its primary benefit is **dimensionality reduction**. By collapsing different forms of a word into a single stem, it drastically reduces the number of unique words (the vocabulary size) in a text corpus, which in turn reduces the memory and processing requirements for subsequent models (Jivani, 2011). However, this efficiency comes at the cost of linguistic precision. Stemming algorithms suffer from two main types of errors:

- **Over-stemming:** This occurs when too much of a word is removed, leading to the incorrect conflation of distinct words (e.g., 'university' and 'universe' might both be stemmed to 'univers').
- **Under-stemming:** This occurs when related words are not reduced to the same stem (e.g., 'adhesion' and 'adhesive').

Furthermore, stemming frequently produces non-dictionary words (e.g., 'comput' from 'computation'), making results difficult to interpret, and operates without any contextual understanding. It represents a classic trade-off: sacrificing linguistic precision for gains in speed and reduced data complexity (Manning & Schütze, 1999).

A.3.3. Current Applications and Use Cases

Despite its crudeness, stemming remains useful in specific applications:

- **Information Retrieval and Search Indexing:** This is a classic use case. Stemming allows a search engine to match a query for "running shoes" with documents containing 'run' or 'ran,' improving recall (Manning et al., 2008).
- **Text Classification Baselines:** In tasks like topic classification or spam detection, where the presence of certain keywords is more important than grammatical nuance, stemming provides a fast and effective way to build baseline models (Vijayarani et al., 2015).

A.4. Lemmatization: Contextual Word Normalization

A.4.1. Evolution of Lemmatization Approaches

Lemmatization is a more sophisticated normalization technique that reduces a word to its base or dictionary form, known as the **lemma**. Unlike stemming, lemmatization guarantees that its output will be a valid, meaningful dictionary word (e.g., 'better' becomes 'good') (Jurafsky & Martin, 2023).

To achieve this, lemmatization must consider the word's context. The process typically involves two critical steps:

1. **Part-of-Speech (POS) Tagging:** The algorithm first determines the grammatical category of the word (e.g., noun, verb, adjective). This is crucial because the lemma can change depending on its POS (e.g., 'meeting' as a noun vs. 'meeting' as a verb).
2. **Dictionary Lookup:** Using the word and its POS tag, the algorithm consults a morphological dictionary (like WordNet) to find the correct lemma (Manning & Schütze, 1999).

A.4.2. Computational Cost and Accuracy Trade-offs

The primary advantage of lemmatization is its linguistic accuracy. By producing valid dictionary words, it improves the interpretability of results and the performance of downstream tasks that rely on semantic understanding (Jivani, 2011).

However, this accuracy comes at a significant computational cost. Lemmatization is inherently more complex and computationally intensive than stemming. The need for POS tagging and dictionary lookups makes the process slower and more resource-heavy, which can be a bottleneck in real-time applications. Its effectiveness is also highly dependent on the availability of comprehensive, language-specific resources (Jurafsky & Martin, 2023). The choice between the two is therefore a clear trade-off between computational efficiency and linguistic accuracy.

A.4.3. Current Applications and Usage

Lemmatization is the preferred normalization technique in applications where semantic meaning is critical:

- **Advanced NLP Tasks:** It is standard in tasks like machine translation, question answering, and conversational AI, where understanding the precise meaning of words is essential (Jurafsky & Martin, 2023).
- **Information Extraction:** Lemmatization helps consolidate extracted entities and relationships, ensuring that different forms of the same concept are grouped together.

- **Feature for Machine Learning:** It is commonly used when preparing text features for complex machine learning models that can leverage semantic information.

A.5. Stop Words Removal: Filtering the Noise

A.5.1. Evolution of Stop Word Handling

Stop words are extremely common words that are often considered to have little semantic value on their own (e.g., 'a,' 'the,' 'is,' 'in'). Stop word removal is the process of filtering these words out of a text corpus. The concept grew out of early work in information retrieval, where researchers observed that removing high-frequency words could improve the performance of search systems (Luhn, 1958).

Initially, stop word removal relied on static, predefined lists. However, a one-size-fits-all approach has limitations, as a word that is a stop word in one context might be a key term in another (e.g., 'will' in a legal document). Modern approaches often involve using domain-specific stop word lists or data-driven methods like **TF-IDF (term frequency-inverse document frequency)** to automatically identify and down-weight non-informative words (Kaur & Saini, 2016).

A.5.2. Impact on Computational Efficiency and Performance

The main benefits of removing stop words are computational:

- **Reduced Dimensionality:** It significantly reduces the size of the vocabulary and the dimensionality of the feature space, which is beneficial for many machine learning algorithms (Manning et al., 2008).
- **Faster Processing:** With less data to process, NLP algorithms can run much faster.
- **Improved Model Performance:** By removing 'noise,' stop word removal can allow models to focus on more meaningful words, which can lead to improved accuracy in tasks like text classification and topic modeling (Vijayarani et al., 2015).

A.5.3. Current Practices and Rationale for Use

Stop word removal remains a common pre-processing step, particularly for tasks where word frequency is a key feature. However, the decision is highly context-dependent. For modern transformer-based models and tasks that rely on word order or syntax—such as machine translation or sentiment analysis of short phrases (e.g., "to be or not to be")—removing stop words is often detrimental as it discards crucial contextual information (Jurafsky & Martin, 2023).

A.6. Bag of Words (BoW): A Foundational Representation

A.6.1. Historical Origins and Development of BoW

After normalization, text must be converted into a numerical format. The **bag-of-words (BoW)** model is one of the earliest and simplest methods for this vectorization. It represents a document as an unordered collection of its words, disregarding all grammar and word order, and focusing only on word counts (Harris, 1954). The process involves tokenizing the text, creating a vocabulary of all unique words in the corpus, and then representing each document as a numerical vector where each element corresponds to a word in the vocabulary and its value is the word's frequency (Manning & Schütze, 1999).

A.6.2. Impact on Computational Processing

The main advantage of BoW is its simplicity and computational efficiency. It provides a straightforward way to convert text into a fixed-length numerical format that can be used by traditional machine learning algorithms (Aggarwal & Zhai, 2012). However, its simplicity is also its greatest weakness:

1. **Loss of Context and Order:** By ignoring word order, BoW loses all syntactic and semantic information encoded in the sequence of words.
2. **High Dimensionality and Sparsity:** For large corpora, the vocabulary can be enormous, leading to very high-dimensional and sparse vectors (mostly filled with zeros), which can be computationally challenging.

These limitations directly motivated the development of more advanced representations. An early improvement was **TF-IDF (term frequency-inverse**

document frequency), which weights the word counts by their relative importance in the corpus (Jones, 1972). A more significant evolutionary leap came with **word embeddings** (e.g., Word2Vec, GloVe), which represent words as dense, low-dimensional vectors that capture rich semantic relationships, overcoming the core limitations of the BoW model (Mikolov et al., 2013; Pennington et al., 2014).

A.6.3. Current Applications and Use Cases

Despite the rise of word embeddings, the BoW model (often in its TF-IDF form) remains a relevant and useful technique:

- **Text Classification and Sentiment Analysis:** BoW is a strong and often sufficient baseline for document classification tasks where the presence of certain keywords is a strong signal (Aggarwal & Zhai, 2012).
- **Topic Modeling:** Algorithms like latent Dirichlet allocation (LDA) are built upon a BoW representation to discover topics within a corpus (Blei et al., 2003).

A.7. Conclusion

The pre-processing techniques detailed in this appendix form the foundational layer upon which nearly all NLP systems are built. Their evolution reflects the broader trajectory of the field, driven by a constant interplay between the need for linguistic accuracy, the constraints of computational resources, and the exponential growth in available text data. Tokenization has matured from simple splitting to complex subword generation to handle linguistic diversity. Stemming and lemmatization offer a classic study in the trade-off between speed and precision. Stop word removal provides significant computational benefits by reducing noise, while the bag-of-words model laid the essential groundwork for text representation, with its limitations directly paving the way for the transformative word embeddings that power modern NLP. Understanding the history, computational impact, and appropriate application of these methods is an indispensable skill for any NLP practitioner.

A.8 References

Aggarwal, C. C., & Zhai, C. (2012). A survey of text classification algorithms. In C. C. Aggarwal & C. Zhai (Eds.), *Mining text data* (pp. 163–222). Springer.

https://doi.org/10.1007/978-1-4614-3223-4_6

Blei, D. M., Ng, A. Y., & Jordan, M. I. (2003). Latent Dirichlet allocation. *Journal of Machine Learning Research, 3*, 993–1022.

Devlin, J., Chang, M. W., Lee, K., & Toutanova, K. (2019). BERT: Pre-training of deep bidirectional transformers for language understanding. *In Proceedings of the 2019 Conference of the North American Chapter of the Association for Computational Linguistics: Human Language Technologies*, Volume 1 (Long and Short Papers) (pp. 4171–4186). Association for Computational Linguistics. https://doi.org/10.18653/v1/N19-1423

Gimpel, K., Schneider, N., O'Connor, B., Das, D., Mills, D., Eisenstein, J., Heilman, M., Yogatama, D., Flanigan, J., & Smith, N. A. (2011). Part-of-speech tagging for Twitter: Annotation, features, and experiments. *In Proceedings of the 49th Annual Meeting of the Association for Computational Linguistics: Human Language Technologies* (pp. 42–47). Association for Computational Linguistics.

Harris, Z. S. (1954). Distributional structure. *Word, 10*(2-3), 146–162. https://doi.org/10.1080/00437956.1954.11659520

Jivani, A. G. (2011). A comparative study of stemming algorithms. *International Journal of Computer Technology and Applications, 2*(6), 1930–1938.

Jones, K. S. (1972). A statistical interpretation of term specificity and its application in retrieval. *Journal of Documentation, 28*(1), 11–21. https://doi.org/10.1108/eb026526

Jurafsky, D., & Martin, J. H. (2023). *Speech and language processing* (3rd ed.). Prentice Hall.

Kaur, J., & Saini, J. R. (2016). A survey on stopword removal algorithms. *International Journal of Computer Applications, 139*(11), 15–18.

Kudo, T., & Richardson, J. (2018). SentencePiece: A simple and language independent subword tokenizer and detokenizer for neural text processing. *In Proceedings of the 2018 Conference on Empirical Methods in Natural Language Processing: System Demonstrations* (pp. 66–71). Association for Computational Linguistics. https://doi.org/10.18653/v1/D18-2012

Lovins, J. B. (1968). Development of a stemming algorithm. *Mechanical Translation and Computational Linguistics, 11*(1/2), 22–31.

Luhn, H. P. (1958). The automatic creation of literature abstracts. *IBM Journal of Research and Development, 2*(2), 159–165. https://doi.org/10.1147/rd.22.0159

Manning, C. D., & Schütze, H. (1999). *Foundations of statistical natural language processing.* MIT Press.

Manning, C. D., Raghavan, P., & Schütze, H. (2008). *Introduction to information retrieval.* Cambridge University Press. https://doi.org/10.1017/CBO9780511809071

Mikolov, T., Chen, K., Corrado, G., & Dean, J. (2013). *Efficient estimation of word representations in vector space.* arXiv preprint arXiv:1301.3781.

Pennington, J., Socher, R., & Manning, C. D. (2014). GloVe: Global vectors for word representation. *In Proceedings of the 2014 Conference on Empirical Methods in Natural Language Processing (EMNLP)* (pp. 1532–1543). Association for Computational Linguistics. https://doi.org/10.3115/v1/D14-1162

Porter, M. F. (1980). An algorithm for suffix stripping. *Program, 14*(3), 130–137. https://doi.org/10.1108/eb046814

Schuster, M., & Nakajima, K. (2012). Japanese and Korean voice search. In *2012 IEEE International*

Conference on Acoustics, Speech and Signal Processing (ICASSP) (pp. 5149–5152). IEEE. https://doi.org/10.1109/ICASSP.2012.6289079

Sennrich, R., Haddow, B., & Birch, A. (2016). Neural machine translation of rare words with subword units. *In Proceedings of the 54th Annual Meeting of the Association for Computational Linguistics* (Volume 1: Long Papers) (pp. 1715–1725). Association for Computational Linguistics. https://doi.org/10.18653/v1/P16-1162

Vijayarani, S., Ilamathi, M. J., & Nithya, M. (2015). Preprocessing techniques for text mining - an overview. *International Journal of Computer Science & Communication Networks, 5*(1), 7–16.

Appendix B

What It Is: The Chain Rule of Probability

The chain rule allows you to calculate the joint probability of a sequence of events occurring. In simple terms, it's a way to find the probability of several things happening in a specific order.

$$P(W) = P(w_1) \times P(w_2|w_1) \times P(w_3|w_1, w_2) \times \cdots \times P(w_k|w_1, \ldots, w_{k-1})$$

can be expressed more compactly as:

$$P(w_1, w_2, \ldots, w_k) = \prod_{i=1}^{k} P(w_i|w_1, \ldots, w_{i-1})$$

where the product (Π) is taken from i=1 to k.

Breakdown of the Notation

- **W**: Represents the entire sequence of events, $(w_1, w_2, \ldots, w_k)$. In many applications, like natural language processing (NLP), W would be a sentence or phrase, and w_i would be the i-th word.
- **P(W)**: The probability of the entire sequence W occurring. For example, if W = "the cat sat," this is the probability of observing that exact phrase.
- **P(w₁)**: The probability of the first event (or word) w_1 occurring on its own.
- **P(w₂ | w₁)**: This is a **conditional probability**. It reads "the probability of w_2 occurring, **given that** w_1 has already occurred."
- **P(w₃ | w₁, w₂)**: The probability of w_3 occurring, **given that** the sequence w_1, w_2 has already occurred.
- **P(wₖ | w₁, ..., wₖ₋₁)**: The probability of the last event w_k occurring, **given that** the entire preceding sequence w_1 through w_{k-1} has occurred.

Intuitive Explanation with an Example

The core idea is: "The probability of a series of things happening is the probability of the first thing, times the probability of the second thing given the first, times the probability of the third thing given the first two, and so on."

Example: Drawing Cards from a Deck (without replacement)

Let's find the probability of drawing three specific cards in order: an Ace, then a King, then a Queen from a standard 52-card deck.

- W = (Ace, King, Queen)
- w_1 = Ace
- w_2 = King
- w_3 = Queen

Using the chain rule:

$$P(\text{Ace, King, Queen}) = P(\text{Ace}) \times P(\text{King}|\text{Ace}) \times P(\text{Queen}|\text{Ace, King})$$

1. **P(Ace)**: The probability of the first card being an Ace. There are 4 Aces in 52 cards.
   - P(Ace) = 4/52
2. **P(King | Ace)**: The probability of the second card being a King, *given* the first was an Ace. Now there are only 51 cards left, but all 4 Kings are still in the deck.
   - P(King | Ace) = 4/51
3. **P(Queen | Ace, King)**: The probability of the third card being a Queen, *given* the first was an Ace and the second was a King. Now there are 50 cards left, and all 4 Queens are still there.
   - P(Queen | Ace, King) = 4/50

So, the total probability is:

$$P(W) = \frac{4}{52} \times \frac{4}{51} \times \frac{4}{50} \approx 0.00048$$

Application in Natural Language Processing (NLP)

This formula is the theoretical foundation for **language models (LMs)**. In this context:

- W is a sentence.
- w are the words in the sentence.
- P(W) is the probability of that sentence being a valid, natural-sounding sentence in a given language.

For example, a good language model would assign:

- P("the cat sat on the mat") >> a high probability
- P("sat the on mat the cat") >> a very low probability

The Problem: The full chain rule is computationally intractable for modeling natural language. The final term, $P(w_k \mid w_1, ..., w_{k-1})$, requires calculating the conditional probability of a word given a potentially extensive history of preceding words. Because the total number of possible sequences grows exponentially with the length of the history, computational models encounter a severe data sparsity issue commonly known as the curse of dimensionality. Consequently, while the number of histories is mathematically finite, the sheer volume of combinations makes it practically impossible to calculate and store all these probabilities.

The Solution: The Markov Assumption

To make this practical, we make a simplifying assumption called the **Markov assumption**: the probability of the next word depends only on a *limited* number of previous words, not the entire history.

- **Bigram Model (n=2):** We assume the next word only depends on the immediately preceding word.

$$P(w_i|w_1,\ldots,w_{i-1}) \approx P(w_i|w_{i-1})$$

The formula becomes:

$$P(W) \approx P(w_1) \times P(w_2|w_1) \times P(w_3|w_2) \times \ldots$$

Trigram Model (n=3): We assume the next word depends on the previous two words.

$$P(w_i|w_1, \ldots, w_{i-1}) \approx P(w_i|w_{i-2}, w_{i-1})$$

These simplified models, known as **N-gram models**, were the state-of-the-art in language modeling for decades and are still used in some applications today. Modern neural network models (like Transformers in GPT) are essentially much more sophisticated ways of estimating these conditional probabilities $P(w_i \mid \text{history})$ without the strict, fixed-window limitations of n-grams.

Appendix C

Markov Assumption for a Bigram Model

This is the critical simplifying step that makes the chain rule of probability practical for use in many applications, especially in early natural language processing (NLP).

Let's break down exactly what this approximation means and why it's so important.

The Equation and Its Meaning

Original (True) Probability:

$$P(w_i|w_1, \ldots, w_{i-1})$$

- This reads: "The probability of the word w_i, given the entire history of all the words that came before it."

- Example: For the sentence "The cat sat on the mat," the true probability of the word 'sat' is P(sat | The, cat).

The Approximation:

$$\approx P(w_i|w_{i-1})$$

- This reads: "is approximately equal to the probability of the word w_i, given only the single word that came immediately before it."

- Example: Using the approximation, we say P(sat | The, cat) $\approx$ P(sat | cat). We completely ignore the word 'The.'

This assumption is called the **Markov assumption** (specifically, a first-order Markov assumption). It states that the present state (w_i) depends only on the immediately preceding state (w_{i-1}) and is independent of all earlier past states $(w_{i-2}, w_{i-3}, ...)$

Why Make This Approximation?

The original chain rule is theoretically perfect but practically impossible for two main reasons:

1. Computational Complexity / Data Sparsity: To calculate P(sat | The, cat), you would need to have seen the exact phrase "The cat" many, many times in your training data to get a reliable statistic for what word comes next. Now imagine trying to calculate the probability of the 10th word in a sentence. The history $(w_1, ..., w_9)$ is so specific that you have likely never seen it before in your entire dataset. This is known as the curse of dimensionality or the data sparsity problem.

2. Storage: Storing the probabilities for every possible sequence of words is computationally intractable.

The bigram assumption solves this by drastically reducing the 'context' we need to consider. We only need to count and store probabilities for pairs of words (bigrams), which is far more manageable.

How it Transforms the Chain Rule

When you apply this approximation to the full chain rule:

*   True Probability:

$$P(\text{the}) \times P(\text{cat}|\text{the}) \times P(\text{sat}|\text{the, cat})$$

*   Bigram Approximation:

$$P(\text{the}) \times P(\text{cat}|\text{the}) \times P(\text{sat}|\text{cat})$$

To calculate **P(sat | cat)**, a system would just need to count how many times it saw the word 'cat' in a massive text corpus, and what fraction of those times it was followed by the word 'sat.' This is easy to compute:

$$P(\text{sat}|\text{cat}) = \frac{\text{count}(\text{cat sat})}{\text{count}(\text{cat})}$$

The Downside: Loss of Context

The bigram model is a powerful simplification, but its weakness is obvious: **it ignores long-range context.**

Consider the sentence:
"The man who was born in France speaks fluent..."

- A bigram model would predict the next word based only on P(word | fluent). It might predict 'English' or 'German' with a high probability because phrases like "fluent English" are common.

- A human (or a more advanced model) would use the context 'France' from much earlier in the sentence to know that the most likely next word is 'French.' The bigram model is blind to this crucial piece of information.

This limitation led to the development of Trigram models (which look at the two previous words), 4-gram models, and ultimately, neural network models like LSTMs and transformers that can learn to handle much longer-range dependencies.

Appendix D

Maximum Likelihood Estimate (MLE)

It's the most direct and intuitive way to calculate the probability of one word following another based on a given text corpus.

Let's break it down with a simple, concrete example.

The Formula: A Practical Definition

$$P(w_i|w_{i-1}) = \frac{\text{count}(w_{i-1}, w_i)}{\text{count}(w_{i-1})}$$

- The Intuition: The probability of word w_i occurring after word w_{i-1} is estimated by the frequency with which it has happened in the past. It's a very straightforward "what you see is what you get" approach.

- The Numerator: This is the raw count of how many times the specific two-word sequence (the bigram) w_{i-1} , w_i appears in your training text.

- The Denominator: This is the raw count of how many times the single word (the unigram) w_{i-1} appears in the same text. This gives us the total number of opportunities for a word to follow w_{i-1}.

Example Calculation

Let's say our entire 'corpus' (our training text) is the following sentence:

Corpus: the cat sat on the mat

Now, let's calculate the probability of the word 'sat' appearing after the word 'cat.'

- w_{i-1} = 'cat'

- w_i = 'sat'

- We want to find: P(sat | cat)

Step 1: Calculate the Numerator
Scan the corpus and count how many times you see the exact sequence "cat sat."
It appears 1 time.

So, count(cat, sat) = 1

Step 2: Calculate the Denominator
Scan the corpus and count how many times you see the word 'cat.'
It appears **1** time.

So, count(cat) = 1

Step 3: Divide

$$P(\text{sat}|\text{cat}) = \frac{\text{count}(\text{cat,sat})}{\text{count}(\text{cat})} = \frac{1}{1} = 1.0$$

Interpretation: Based on our tiny corpus, if we see the word 'cat,' the probability of the next word being 'sat' is 100%.

The Major Problem with this Simple Formula: Data Sparsity

This MLE approach has a critical weakness: the **zero-frequency problem**.

What is the probability P(dog | the) according to our corpus?

1. Numerator The sequence "the dog" never appears. The count is 0.

2. **Denominator** The word 'the' appears two times.

3. **Divide:**

$$P(\text{dog|the}) = \frac{0}{2} = 0$$

The model assigns a probability of **zero** to a perfectly reasonable phrase just because it wasn't in the limited training data. If a language model is calculating the probability of a sentence like "the dog barked," the moment it hits the P(dog | the) term, the entire sentence probability becomes zero (... * 0 * ... = 0), making the model extremely brittle.

The Solution: Smoothing

To solve this, we use techniques called **smoothing** (or **additive smoothing**). The simplest form is **Laplace Smoothing** (or Add-1 smoothing), which modifies the formula like this:

$$P(w_i|w_{i-1}) = \frac{\text{count}(w_{i-1}, w_i) + 1}{\text{count}(w_{i-1}) + V}$$

Where **V** is the size of the vocabulary (the number of unique words in the corpus). This technique effectively 'borrows' a tiny bit of probability mass from the things we've seen and gives it to the things we haven't, ensuring that no probability is ever truly zero. See Appendix E for a detailed explanation of Laplace Smoothing.

Appendix E

Laplace Smoothing (also known as Add-1 Smoothing)

This is a fundamental and widely-taught technique used to solve the critical **zero-frequency problem** that arises from the simple Maximum Likelihood Estimate (MLE) formula we discussed in Appendix D.

Let's do a deep dive into this formula.

The Formula

$$P_{\text{Laplace}}(w_i|w_{i-1}) = \frac{\text{count}(w_{i-1}, w_i) + 1}{\text{count}(w_{i-1}) + V}$$

The Problem It Solves: Zero Probabilities

Let's recall the simple MLE formula: count(bigram) / count(unigram). Imagine our training corpus is the cat sat on the mat.

- P(cat | the) = 1/2

- P(dog | the) = 0/2 = 0

A model using this formula would conclude that the phrase "the dog" is **impossible**. If it tried to calculate the probability of the sentence "the dog barked," the entire probability would become zero, because you would multiply by P(dog | the).

This makes the model extremely brittle and unable to handle perfectly valid word combinations that just didn't happen to be in its training data.

The Solution: Laplace Smoothing

Example

Corpus: the cat sat on the mat
Vocabulary (V): {the, cat, sat, on, mat}, so V=5.

Let's calculate P(cat | the):

- MLE (No Smoothing)

$$P(\text{cat}|\text{the}) = \frac{\text{count}(\text{the,cat})}{\text{count}(\text{the})} = \frac{1}{2} = 0.5$$

- Laplace (Add-1) Smoothing

$$P_{\text{Laplace}}(\text{cat}|\text{the}) = \frac{\text{count}(\text{the,cat}) + 1}{\text{count}(\text{the}) + V} = \frac{1+1}{2+5} = \frac{2}{7} \approx 0.286$$

Notice the probability went down. We 'stole' some of the probability mass from the event we saw ("the cat") to redistribute it among events we didn't see.

Now, let's calculate P(sat | the) instead, which is an unseen bigram or a zero-count event.

- MLE (No Smoothing)

$$P(\text{sat}|\text{the}) = \frac{\text{count}(\text{the,sat})}{\text{count}(\text{the})} = \frac{0}{2} = 0$$

- Laplace (Add-1) Smoothing

$$P_{\text{Laplace}}(\text{sat}|\text{the}) = \frac{\text{count}(\text{the,sat}) + 1}{\text{count}(\text{the}) + V} = \frac{0+1}{2+5} = \frac{1}{7} \approx 0.143$$

We now have a non-zero probability for an unseen bigram, making our model far more **robust**.

Weakness of Laplace Smoothing

While simple and effective, Laplace smoothing has a major drawback: **it gives too much probability mass to unseen events.**

In a real-world scenario with a vocabulary of 50,000 ($V = 50{,}000$), the denominator $\text{count}(w_{i-1}) + 50000$ becomes huge. This heavily discounts the counts of all the frequent, observed bigrams. The model becomes "too smooth" and doesn't trust the data it actually saw.

For this reason, more sophisticated smoothing techniques are used in practice, such as:

- Add-k Smoothing: A generalization where you add a fraction k instead of 1.

- Good-Turing Smoothing: A more complex statistical method that estimates how much probability to assign to unseen events based on the frequency of events seen only once.

- Kneser-Ney Smoothing: Often considered the state-of-the-art for traditional n-gram models, it uses a more clever way to "back off" to lower-order n-gram probabilities. This means that if a higher-order n-gram (e.g., a trigram) hasn't been seen often enough, the model relies on the probability of the lower-order n-gram (e.g., the bigram) as a fallback estimate. Kneser-Ney performs this backing off in a particularly effective way, considering the diversity of contexts a word appears in.

Appendix F

The Transformer Architecture: A Technical Deep Dive

F.1 Introduction

The main body of Chapter 13 provides a high-level, conceptual overview of the Transformer architecture and its significance. This appendix offers a more comprehensive technical review for readers interested in the underlying mechanics. We will deconstruct the mathematical foundations of Scaled Dot-Product Attention, provide a more detailed analysis of the model's components, and explore the architectural limitations that spurred the development of more efficient successor models.

F.2 Deconstructing Scaled Dot-Product Attention

The core mechanism of the Transformer is its self-attention function. The official formula, as presented by Vaswani et al. (2017), is:

Equation F.1

Scaled Dot-Product Attention

$$\text{Attention}(Q, K, V) = \text{softmax}\left(\frac{QK^T}{\sqrt{d_k}}\right) V$$

Figure F.1

Scaled Dot-Product Attention

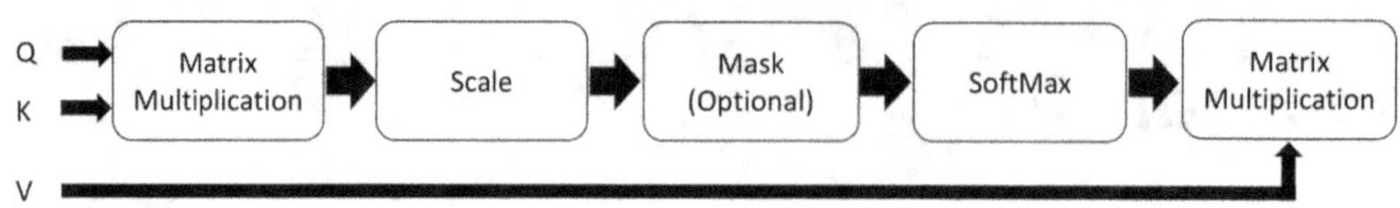

Here, Q, K, and V represent the matrices containing the Query, Key, and Value vectors for all tokens in the sequence, respectively. The term d_k is the dimension of the key vectors. A step-by-step breakdown illuminates this process (Vaswani et al., 2017):

1. **Input Matrix Creation:** The process begins with an input matrix X of dimension $[n \times d_{model}]$, where n is the sequence length and d_{model} is the embedding dimension. This input is projected into Q, K, and V matrices by multiplying it with learned weight matrices W^Q, W^K, and W^V. The resulting matrices are $Q = XW^Q$, $K = XW^K$, and $V = XW^V$.

2. **Dot-Product Similarity Score Calculation:** The first operation is the matrix multiplication QK^T. The resulting matrix has dimensions $[n \times n]$. Each element (i, j) of this matrix contains the dot product of the query vector for the *i-th* token and the key vector for the *j-th* token, serving as a measure of compatibility between them (Vaswani et al., 2017).

3. **Scaling for Stability:** The entire $[n \times n]$ score matrix is then scaled by dividing every element by d_k. This step is critical for stable training. The authors of the original paper noted that for large values of d_k, the magnitude of the dot products can become very large, pushing the softmax function into regions where its gradient is infinitesimally small. Scaling mitigates this by keeping the variance of the inputs to the softmax function controlled (Vaswani et al., 2017).

4. **Softmax for Attention Weights:** A softmax function is applied row-wise to the scaled score matrix. This converts the raw compatibility scores into a probability distribution of attention weights, where each row sums to 1. Each weight indicates the amount of attention the *i-th* token should pay to the *j-th* token (Vaswani et al., 2017).

5. **Weighted Sum of Values:** Finally, the attention weight matrix is multiplied

by the Value matrix V. The result is an output matrix where each row i is a new, contextually enriched representation for the i-th token, computed as the weighted sum of all the Value vectors in the sequence (Vaswani et al., 2017).

About the Masking Mechanism

Masking is the process of hiding or ignoring certain parts of the input from the model's attention mechanism (Vaswani et al., 2017). The 'optional' nature of this step in the attention diagram signifies that it is not always used. It is specifically applied in the decoder's self-attention sub-layer to prevent positions from attending to subsequent positions. This is crucial for autoregressive generation tasks, as it ensures that the prediction for a token at position i can only depend on the known outputs at positions less than i, preventing the model from 'cheating' by looking ahead at the answer during training. The mask is implemented by adding a large negative value (negative infinity) to the attention scores corresponding to future positions before the softmax step, which effectively makes their resulting attention weights zero (Vaswani et al., 2017).

F.3 Architectural Details: A Layer-by-Layer Analysis

F.3.1 Multi-Head Attention

Instead of performing a single attention function, the Transformer employs Multi-Head Attention, which allows the model to jointly attend to information from different representation subspaces at different positions. This is accomplished by running the Scaled Dot-Product Attention mechanism multiple times in parallel (Vaswani et al., 2017):

Equation F.2

Multi-Head Attention (Note: From Vaswani et al., 2017)

$$\text{MultiHead}(Q, K, V) = \text{Concat}(\text{head}_1, \ldots, \text{head}_h)W^O$$

$$\text{where head}_i = \text{Attention}(QW_i^Q, KW_i^K, VW_i^V)$$

1. **Linear Projection:** Multi-Head Attention uses h different sets of learned projection matrices (W_i^Q, W_i^K, and W_i^V for each head$_i = 1 \ldots h$)

    These matrices project the input into lower-dimensional subspaces. Typically, the dimension for each head is set to $d_k = d_v = d_{model}/h$ (Vaswani et al., 2017).

2. **Parallel Attention Calculation:** Scaled Dot-Product Attention is then performed independently and in parallel for each of the h heads, resulting in h separate output matrices (Vaswani et al., 2017).

3. **Concatenation and Final Projection:** The h output matrices are concatenated and passed through one final, learned linear projection matrix, W^O, to produce the final output.

F.3.2 The "Add & Norm" Sub-layer

Figure F.2

Structure of a Transformer Encoder Layer

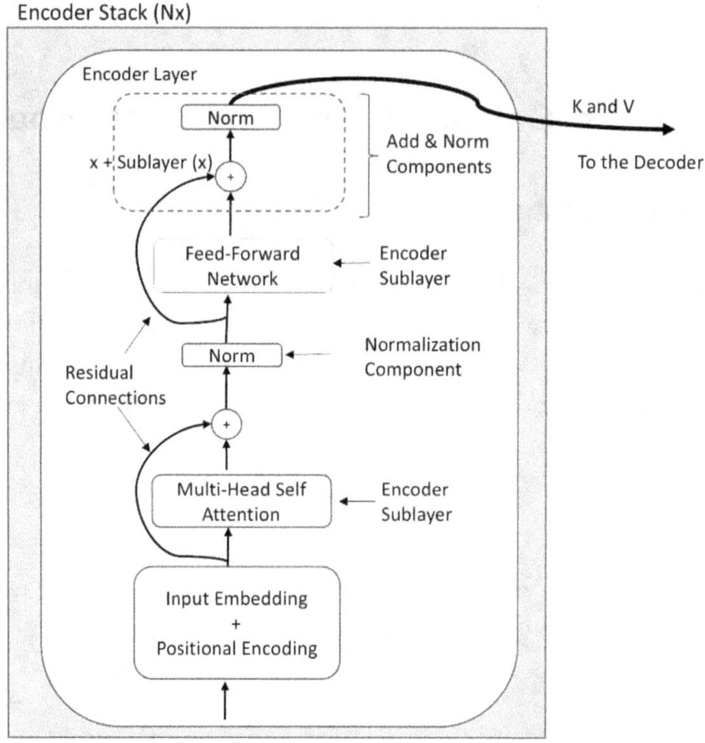

The "Add & Norm" component, which follows each sub-layer (Multi-Head Attention and the Feed-Forward Network), is a fusion of two critical techniques for training deep networks:

1. **Residual Connection:** The output of a sub-layer is added to the sub-layer's original input: x + Sublayer(x). This provides a direct path for the gradient to flow during backpropagation, which is crucial for alleviating the vanishing gradient problem and enabling the stable training of very deep networks (He et al., 2016).

2. **Layer Normalization:** Following the addition, the result is passed through a Layer Normalization step. Layer Normalization computes normalization statistics (mean and variance) independently for each sequence element across its feature dimension. This stabilizes the hidden state dynamics and makes the training process faster and more reliable, especially for variable-length sequences (Ba et al., 2016).

F.4 The Quadratic Bottleneck: Analyzing the O(n²) Complexity

The most significant architectural limitation of the original Transformer is that its self-attention operation requires computational time and memory that scales quadratically with the sequence length, n, denoted as $O(n^2)$. This arises because the QK^T matrix multiplication computes a similarity score between every pair of tokens in the sequence, resulting in an n $\times$ n attention matrix (Vaswani et al., 2017).

About O(n²)

In computer science, Big O notation is used to describe the performance or complexity of an algorithm as the size of its input grows. $O(n^2)$ (read as "Order of n-squared") signifies that the algorithm's execution time or memory requirement will grow quadratically with the size of the input, n (Cormen et al., 2022). For the Transformer, n is the sequence length. This means that if you double the length of the input sequence, the computational cost of the self-attention layer will quadruple. This quadratic scaling makes processing very long sequences, such as entire documents or books, computationally infeasible with the standard Transformer architecture.

The practical implications of this quadratic bottleneck are severe. It forced early practitioners to truncate inputs to a fixed length, potentially discarding vital information from longer documents (Devlin et al., 2019). This limitation became a primary driver for a new wave of research aimed at creating more efficient Transformer variants that could handle longer sequences.

F.5 Evolving the Transformer: A Survey of Efficient Attention Mechanisms

To address the quadratic bottleneck, researchers have proposed numerous "efficient attention" mechanisms. Two of the most influential are the longformer and the reformer.

F.5.1. The Longformer: Linear Scaling with Local and Global Attention

The Longformer model addresses the quadratic complexity by using a sparse attention pattern that scales linearly with sequence length, $O(n)$. It achieves this by combining two types of attention: a sliding window (local) attention, where each token attends to a fixed window of neighboring tokens, and a global attention mechanism, where a few pre-selected tokens (like the special [CLS] token) are allowed to attend to the entire sequence. This hybrid approach allows the model to process thousands of tokens while still capturing both local context and task-critical global information (Beltagy et al., 2020).

About the [CLS] Token

[CLS] stands for 'Classification.' It is a special token that is added to the beginning of every input sequence before it is fed into a Transformer model like BERT. During the self-attention process, the representation for the [CLS] token is built by attending to all the other words in the sentence. As a result, its final hidden state is used as an aggregate sequence representation for classification tasks. This single vector is then typically passed to a simple classifier to make a prediction for the entire sequence (e.g., 'positive' or 'negative') (Devlin et al., 2019).

F.5.2. The Reformer: Efficiency via Hashing and Reversibility

The Reformer introduced two key innovations to improve efficiency. First, it replaces standard dot-product attention with **locality-sensitive hashing (LSH) attention**.

Instead of comparing every Query with every Key (which takes $O(n^2)$ time), LSH attention uses a clever hashing technique as a shortcut. Imagine sorting all the Key vectors into different 'buckets' (shelves in a library) such that similar Keys are highly likely to end up in the same bucket. When a Query vector arrives, the LSH function quickly identifies the most likely bucket containing relevant Keys. The Query then only needs to compare itself against the Keys within that single bucket, drastically reducing the number of comparisons. This approximation changes the complexity from $O(n^2)$ to $O(n \log n)$, making it much faster for long sequences.

Second, it uses **reversible residual layers**. These specially designed layers allow the **activations** (the outputs computed by a layer) from any given layer to be precisely recalculated during the backward pass of training using only the activations of the *next* layer. Because standard training requires storing activations from all layers in memory to compute gradients, this reversibility means the Reformer can avoid storing most activations, significantly reducing memory usage. This innovation makes the memory required for training largely independent of the number of layers, enabling the creation of extremely deep and memory-efficient models (Kitaev et al., 2020).

F.6 References

Ba, J. L., Kiros, J. R., & Hinton, G. E. (2016). *Layer normalization.* arXiv preprint arXiv:1607.06450.

Beltagy, I., Peters, M. E., & Cohan, A. (2020). *Longformer: The long-document transformer.* arXiv preprint arXiv:2004.05150.

Cormen, T. H., Leiserson, C. E., Rivest, R. L., & Stein, C. (2022). *Introduction to algorithms* (4th ed.). MIT Press.

Devlin, J., Chang, M. W., Lee, K., & Toutanova, K. (2019). BERT: Pre-training of deep bidirectional transformers for language understanding. *In Proceedings of the 2019 Conference of the North American Chapter of the Association for Computational Linguistics: Human Language Technologies*, Volume 1 (Long and Short Papers) (pp. 4171–4186). Association for Computational Linguistics. https://doi.org/10.18653/v1/N19-1423

He, K., Zhang, X., Ren, S., & Sun, J. (2016). Deep residual learning for image recognition. In *Proceedings of the Ieee Conference on Computer Vision and Pattern Recognition* (pp. 770–778). https://doi.org/10.1109/CVPR.2016.90

Kitaev, N., Kaiser, Ł., & Levskaya, A. (2020). Reformer: The efficient transformer. In *International Conference on Learning Representations.*

Vaswani, A., Shazeer, N., Parmar, N., Uszkoreit, J., Jones, L., Gomez, A. N., Kaiser, Ł., & Polosukhin, I. (2017). Attention is all you need. In *Advances in neural information processing systems, 30.*

Appendix G

The Evolution of NLP Paradigms: A Comparative Summary

Introduction

The history of natural language processing (NLP) is a story of paradigm shifts, with each era of innovation building upon the successes and addressing the limitations of the last. This appendix provides a consolidated, comparative summary of the major technical paradigms discussed throughout this book. It traces the evolution of the field from early rule-based systems to the modern deep learning era, highlighting the key techniques, computational trade-offs, and core applications that have defined each stage of development. This summary is intended to serve as a quick-reference guide for understanding the historical context and foundational principles that have shaped modern artificial intelligence.

Table G.1

Comparative Summary of NLP Paradigms

Paradigm / Era	Key Techniques & Models	Core Idea & Theoretical Foundation	Computational Impact & Limitations	Key Applications
1950s–1980s: Rule-Based Systems	• Handcrafted grammars (e.g., CFGs) • Ontologies & Knowledge Bases • Expert Systems	• Language can be understood by explicitly encoding linguistic rules and world knowledge • Relies heavily on human expertise from linguists	**Impact:** Early successes in constrained domains (e.g., SHRDLU). **Limitations:** Extremely brittle; failed to handle ambiguity or variation. Required immense manual effort, making them domain-specific, and not scalable or adaptable to new data.	• Early machine translation • Question answering in closed domains • Natural language interfaces to databases
1980s–2000s: Statistical Methods	• N-grams • Hidden Markov Models (HMMs) • Probabilistic Parsing	• Language patterns and grammatical structure can be learned automatically from large text corpora • Uses probability theory to handle ambiguity	**Impact:** Shifted the burden from manual rule-writing to data collection. **Limitations:** Suffered from severe data sparsity (the "zero-frequency problem"). Required smoothing techniques. HMMs had restrictive independence assumptions and struggled to capture long-range dependencies.	• Predictive text and speech recognition. •Part-of-Speech (POS) tagging • Early statistical machine translation (SMT)

Paradigm / Era	Key Techniques & Models	Core Idea & Theoretical Foundation	Computational Impact & Limitations	Key Applications
2000s–2010s: Classic Machine Learning	• Bag-of-Words (BoW) & TF-IDF • Models: Naive Bayes, Support Vector Machines (SVMs), Logistic Regression, Conditional Random Fields (CRFs)	• Framed NLP tasks as classification or sequence labeling problems • Required extensive manual feature engineering to convert text into informative numerical vectors for the models	**Impact**: Achieved strong performance on many classification tasks (e.g., spam detection, sentiment analysis). CRFs excelled at sequence labeling (e.g., NER). **Limitations**: Performance heavily dependent on feature engineering. Models had no inherent understanding of semantics (e.g., 'car' and 'automobile' are unrelated symbols).	• Text classification (spam detection, topic labeling, and sentiment analysis) • Named Entity Recognition (NER)
2010s–Present: Deep Learning Revolution	• Word Embeddings (Word2Vec, GloVe) • Recurrent Neural Networks (RNNs) & LSTMs	•Automatic feature learning: Models learn dense, semantic representations (embeddings) directly from raw text data	**Impact**: State-of-the-art performance on nearly all NLP tasks. Eliminated the need for manual feature engineering.	• Modern machine translation • Question answering • Text summarization

Paradigm / Era	Key Techniques & Models	Core Idea & Theoretical Foundation	Computational Impact & Limitations	Key Applications
2010s–Present: Deep Learning Revolution	•Sequence-to-Sequence (Seq2Seq) with Attention • Transformers (BERT, GPT series)	• Architectural innovations (e.g., gates in LSTMs, self-attention in Transformers) designed to capture long-range dependencies	Parallelization in Transformers enabled massive scaling. **Limitations**: Requires massive datasets and immense computational resources (GPUs). Models are often "black boxes" that are difficult to interpret. Prone to issues like bias and hallucination.	• Conversational AI (chatbots) • Large language models (LLMs) with broad capabilities

Appendix H

Extended Practical Case Studies

Case Study 1: Sentiment Analysis at Scale

Objective
Build a real-time sentiment analysis system for brand monitoring on Twitter.

Background & Motivation
Social media sentiment drives customer engagement and brand reputation, informing marketing strategies.

Problem Statement
Detect positive, negative, and neutral sentiment for a particular brand across live tweet streams.

Data Pipeline

Collection: Stream tweets via the Twitter API filtered by brand keywords.

Preprocessing: Tokenize text, remove URLs/mentions/emojis, lowercase, stopword removal.

Storage & Management: Store JSON in MongoDB for horizontal scalability.

Methodology & Modeling
Fine-tune BERT (bert-base-uncased) on an SST-2–style sentiment dataset, adjusting learning rate (2e-5) and batch size (16).

Code Snippets

```python
from transformers import BertForSequenceClassification, Trainer,
TrainingArguments

# Load model and tokenizer
model = BertForSequenceClassification.from_pretrained('bert-base-uncased',
num_labels=3)

# Training setup
...
```

Evaluation & Results

- Metrics: 92.3% accuracy, F1-score of 0.91

- Analysis: Confusion matrix revealed misclassifications mainly between neutral and positive classes

Lessons Learned

- Real-time inference demands efficient batching to reduce latency

- GPU acceleration yields 3× faster throughput compared to CPU

Extensions & "What If?"

- What if we substitute DistilBERT for faster inference at slight accuracy loss?

- How would multilingual support affect preprocessing and model choice?

Case Study 2: Legal Document Summarization

Objective
Automate concise summarization of court opinions to expedite legal research.

Background & Motivation
Lawyers and researchers spend hours reviewing lengthy rulings; automated summaries increase efficiency.

Problem Statement
Generate abstractive summaries preserving critical facts and holdings from legal

texts.

Data Pipeline

- Collection: Download public court opinions via PACER/open legal corpora

- Preprocessing: Segment sentences, normalize citations, remove boilerplate

- Storage & Management: Save as plain text or JSON annotations

Methodology & Modeling

Compare prompt-based GPT-3.5 summarization vs. a fine-tuned T5 model (t5-base) on legal summary pairs.

Code Snippets

```python
from transformers import T5ForConditionalGeneration,
T5Tokenizer

# Initialize model
model = T5ForConditionalGeneration.from_pretrained('t5-
base')

# Fine-tuning loop
...
```

Evaluation & Results

- Metrics: ROUGE-1 45.8, ROUGE-2 24.1; human scores: Informativeness 4.2/5, Fluency 4.0/5

- Analysis: GPT-3.5 prompts yielded more fluent text but occasional factual omissions

Lessons Learned

- Extractive Vs. Abstractive Trade-Offs: abstraction improves readability but risks hallucination

- Human-in-the-loop review reduces critical errors

Extensions & "What If?"

- Test extractive baselines (TextRank) for faster deployment

- Evaluate impact of legal citation formats on summary quality

Case Study 3: Customer Support Chatbot

Objective
Develop a conversational agent to handle tier-1 customer inquiries and escalate when needed.

Background & Motivation
Automating FAQs reduces support costs and improves response times.

Problem Statement
Create a chatbot that classifies intents, extracts entities, and provides accurate answers or hands off to an agent.

Data Pipeline

- Collection: Historical chat logs and FAQ documents.

- Preprocessing: Label intents/entities; normalize user utterances

- Storage & Management: Use YAML/JSON for intent definitions and NLU data

Methodology & Modeling
Use Rasa's DIETClassifier for NLU and implement Retrieval-Augmented Generation (RAG) fallback.

Implementation Details

```
pipeline:
  - name: WhitespaceTokenizer
  - name: CountVectorsFeaturizer
  - name: DIETClassifier
policies:
  - name: MemoizationPolicy
  - name: TEDPolicy
```

Evaluation & Results

- Metrics: Intent accuracy 89.5%, resolution rate 78%, user satisfaction 4.1/5

- Analysis: RAG improved fallback relevance by 15% over static responses

Lessons Learned

- Balancing scripted dialogues with generative fallbacks to maintain coherence

- Continuous retraining on real-world logs improves performance

Extensions & "What If?"

- Integrate sentiment detection to tailor responses

- Implement live agent handoff with context transfer

References & Tools

- Hugging Face Transformers (2023)

- Rasa Open Source (2023)

- Twitter API Documentation (2023)

Usage Note: Readers can follow these studies end-to-end, adapting code to their own environments. Familiarity with Python and ML frameworks is assumed.

H.1 References

Bocklisch, T., Faulkner, J., Garafolo, N., & Nichol, M. (2017). *Rasa: Open source language understanding and dialogue management.* arXiv preprint arXiv:1712.05181. https://arxiv.org/abs/1712.05181

Brown, T., Mann, B., Ryder, N., Subbiah, M., Kaplan, J. D., Dhariwal, P., Neelakantan, A., Shyam, P., Sastry, G., Askell, A., Agarwal, S., Herbert-Voss, A., Krueger, G., Henighan, T., Child, R., Ramesh, A., Ziegler, D., Wu, J., Winter, C., ... Amodei, D. (2020). Language models are few-shot learners. In H. Larochelle, M. Ranzato, R. Hadsell, M. F. Balcan, & H. Lin (Eds.), *Advances in Neural Information Processing Systems* 33 (NeurIPS 2020) (pp. 1877–1901). Curran Associates.

Devlin, J., Chang, M.-W., Lee, K., & Toutanova, K. (2019). BERT: Pre-training of deep bidirectional transformers for language understanding. *In Proceedings of the 2019 Conference of the North American Chapter of the Association for Computational Linguistics: Human Language Technologies*, Volume 1 (Long and Short Papers) (pp. 4171–4186). Association for Computational Linguistics. https://doi.org/10.18653/v1/N19-1423

Karpukhin, V., Oguz, B., Min, S., Lewis, P., Wu, L., Edunov, S., Chen, D., & Yih, W. (2020). Dense passage retrieval for open-domain question answering. *In Proceedings of the 2020 Conference on Empirical Methods in Natural Language Processing (EMNLP)* (pp. 6769–6781). Association for Computational Linguistics. https://doi.org/10.18653/v1/2020.emnlp-main.550

Raffel, C., Shazeer, N., Roberts, A., Lee, K., Narang, S., Matena, M., Zhou, Y., Li, W., & Liu, P. J. (2020). Exploring the limits of transfer learning with a unified text-to-text transformer. *Journal of Machine Learning Research*, 21(140), 1–67. http://jmlr.org/papers/v21/20-074.html

Wang, A., Singh, A., Michael, J., Hill, F., Levy, O., & Bowman, S. R. (2018). GLUE: A multi-task benchmark and analysis platform for natural language understanding. *In Proceedings of the 2018 EMNLP Workshop BlackboxNLP: Analyzing and Interpreting Neural Networks for NLP* (pp. 353–355). Association for Computational Linguistics. https://doi.org/10.18653/v1/W18-5446

H.2 Tools & Documentation

Legal Information Institute. (n.d.). Cornell Law School. Retrieved September 5, 2025, from https://www.law.cornell.edu/

MongoDB, Inc. (n.d.). MongoDB manual. Retrieved September 5, 2025, from https://www.mongodb.com/docs/

PACER Service Center. (n.d.). Public Access to Court Electronic Records. Administrative Office of the U.S. Courts. Retrieved September 5, 2025, from https://pacer.uscourts.gov/

Rasa open source documentation. Retrieved September 5, 2025, from https://rasa.com/docs/rasa/

Twitter Developer. (n.d.). Twitter API documentation. Twitter Developer Platform. Retrieved September 5, 2025, from https://developer.twitter.com/en/docs/twitter-api

Appendix I

Multilingual & Low-Resource NLP Techniques

This appendix explores strategies, tools, and benchmarks for tackling NLP tasks in multilingual and low-resource settings. Each section provides conceptual background, implementation guidance, and pointers to resources.

I.1 Cross-Lingual Transfer Learning

Concept: Leveraging pretrained multilingual models (e.g., mBERT, XLM-R) to transfer knowledge across languages.

Techniques:

- **Zero-shot transfer -** Direct inference in a target language without task-specific training data

- **Few-shot adaptation -** Fine-tuning on a small labeled corpus in the target language

Implementation Tips:

- Use Hugging Face's AutoModelForSequenceClassification with multilingual checkpoints

- Maintain vocabulary coverage by merging language tokens

I.2 Data Augmentation Strategies

Concept: Expanding scarce labeled datasets via synthetic or weak supervision.

Methods:

- **Back-translation:** translate text to an auxiliary language and back to generate paraphrases

- **Synthetic data generation**: use language generation models to create labeled examples

Annotation projection: transfer annotations via aligned parallel corpora.

Implementation Tips:

- Leverage MarianMT from Hugging Face for efficient back-translation

- Control noise by filtering via language-model perplexity thresholds

I.3 Unsupervised & Weakly Supervised Methods

Concept: Reducing reliance on labeled data through self-supervision and distant supervision.

Approaches:

Masked language modeling adaptations: pretrain on large unlabeled corpora in target language.

Distant supervision: use existing resources (e.g., Wikipedia links) to infer labels.

Implementation Tips:

- Collect raw text from Common Crawl for pretraining with mBART or RoBERTa variants

- Use heuristics (like bilingual lexicons) to generate silver-standard training labels

I.4 Benchmarking and Evaluation

Concept: Measuring performance across languages and tasks with standardized benchmarks.

Benchmarks:

- XTREME: cross-lingual multimodal evaluation suite

- MasakhaNER: Named Entity Recognition for African languages

Best Practices:

- Report per-language metrics to reveal disparities

- Analyze error patterns by language family and script

I.5 Community & Open Resources

Datasets:

- FLORES-101: multilingual translation evaluation set

- Masakhane corpora: African language datasets

Toolkits:

- Hugging Face Datasets: wide array of multilingual collections

- Indic NLP Library: for Indian languages

Communities:

- Masakhane: community for African NLP

- XLM-R Workshop: workshop series on multilingual representation

References

Artetxe, M., & Schwenk, H. (2019). Massively multilingual sentence embeddings for zero-shot cross-lingual transfer and beyond. Transactions of the Association for Computational Linguistics, 7, 597–610. https://doi.org/10.1162/tacl_a_00288

Conneau, A., Khandelwal, K., Goyal, N., Chaudhary, V., Wenzek, G., Guzmán, F., Grave, E., Ott, M., Zettlemoyer, L., & Stoyanov, V. (2020). Unsupervised cross-lingual representation learning at scale. *In Proceedings of the 58th Annual Meeting of the Association for Computational Linguistics* (pp. 8440–8451). Association for Computational Linguistics. https://doi.org/10.18653/v1/2020.acl-main.747

Edunov, S., Ott, M., Auli, M., & Grangier, D. (2018). Understanding back-translation at scale. *In Proceedings of the 2018 Conference on Empirical Methods in Natural Language Processing* (pp. 489–500). Association for Computational Linguistics. https://doi.org/10.18653/v1/D18-1045

Fadaee, M., Bisazza, A., & Monz, C. (2017). Data augmentation for low-resource neural machine translation. *In Proceedings of the 55th Annual Meeting of the Association for Computational Linguistics* (Volume 2: Short Papers) (pp. 567–573). Association for Computational Linguistics. https://doi.org/10.18653/v1/P17-2090

Hu, J., Ruder, S., Siddhant, A., Neubig, G., Firat, O., & Johnson, M. (2020). XTREME: A

massively multilingual multi-task benchmark for evaluating cross-lingual generalization. In H. Daumé III & A. Singh (Eds.), *Proceedings of the 37th International Conference on Machine Learning* (pp. 4314–4323). PMLR. http://proceedings.mlr.press/v119/hu20b.html

Lewis, M., Liu, Y., Goyal, N., Ghazvininejad, M., Mohamed, A., Levy, O., Stoyanov, V., & Zettlemoyer, L. (2020). BART: Denoising sequence-to-sequence pre-training for natural language generation, translation, and comprehension. *In Proceedings of the 58th Annual Meeting of the Association for Computational Linguistics* (pp. 7871–7880). Association for Computational Linguistics. https://doi.org/10.18653/v1/2020.acl-main.703

Pires, T., Schlinger, E., & Garrette, D. (2019). How multilingual is multilingual BERT? *In Proceedings of the 57th Annual Meeting of the Association for Computational Linguistics* (pp. 4996–5001). Association for Computational Linguistics. https://doi.org/10.18653/v1/P19-1493

Sennrich, R., Haddow, B., & Birch, A. (2016). Improving neural machine translation models with monolingual data. *In Proceedings of the 54th Annual Meeting of the Association for Computational Linguistics* (Volume 1: Long Papers) (pp. 86–96). Association for Computational Linguistics. https://doi.org/10.18653/v1/P16-1009

Index